George Husband Baird Macleod

Notes on the Surgery of the War in the Crimea

With Remarks on the Treatment of Gunshot Wounds

George Husband Macleod

Notes on the Surgery of the War in the Crimea
With Remarks on the Treatment of Gunshot Wounds

ISBN/EAN: 9783337041755

Printed in Europe, USA, Canada, Australia, Japan

Cover: Foto ©berggeist007 / pixelio.de

More available books at **www.hansebooks.com**

NOTES ON THE SURGERY

OF THE

WAR IN THE CRIMEA,

WITH REMARKS ON

THE TREATMENT OF GUNSHOT WOUNDS.

BY

GEORGE H. B. MACLEOD, M.D., F.R.C.S.

Formerly Surgeon to the Civil Hospital at Smyrna, and to the
General Hospital in Camp before Sebastopol:
Lecturer on Military Surgery in Anderson's University,
Glasgow, &c.

J. W. RANDOLPH,
121 MAIN STREET, RICHMOND, VA.
1862.

EDITOR'S PREFACE.

, the republication of this work in its present somewhat abridged form, the Editor has striven, while reducing its bulk within the narrowed resource 'f the times, to incorporate in it all of the more essential and pra ʼal features of the original volume, and thus to present to the field surgeon, in a convenient form, the highest recent authority to which reference could be had.

The want of access to the standard European authors upon Military Surgery has been seriously felt by the Medical Corps of our army, and if this volume contributes anything towards the satisfaction of that want, the Editor will be more than compensated for the labor bestowed upon its reproduction.

A. N. TALLEY, Surgeon P. A. C. S.,
President Army Medical Board
at Richmond, Va.

NOTES

ON THE

SURGERY OF THE CRIMEAN WAR.

CHAPTER I.

Distinction between Surgery as practised in the Army and in Civil Life—Soldiers as Patients, and the Character of the injuries to which they are liable—Some Peculiarities in the Wounds and Injuries seen during the late War.

That military surgery does not differ from the surgery of civil life, is an assertion which is true in letter, but not in spirit. As a science, surgery, wherever practised, is one and indivisible; but as an art, it varies according to the peculiar nature of the injuries with which it has to deal, and with the circumstances in which it falls to be exercised. To the surgeon practising in the camp, many accidents are presented which seldom or never come within the observation of the civil practitioner; while not a few of the cases which are daily treated in domestic life, rarely come under the charge of the military surgeon. The two classes of practitioners may be said to be engaged in separate departments of the same profession, which, though uniting occasionally, are yet tolerably distinct from one another.

The military surgeon during peace enters for a time into civil life; but during war he is called upon to exercise the very highest functions of his profession, and has little to do with the more trivial accidents which constitute the sum of a private practitioner's daily routine. His observation is undoubtedly restricted to a smaller variety of cases. He sees less than the civilian of the modifications which are impressed upon disease by age and sex; but in war he has a wider field for noticing the influence of external circumstances, of extremes of climate, of variations in food, work, and shelter on the same men, as well as the effects of mental causes, as seen in the exultation of victory and in the prostration and dejection of defeat.

But though there may exist such distinctions between the spheres of the military and those of the civil surgeon, there is surely nothing in the exercise of their different callings which should create an antagonism between them. They are both members of the same priesthood, whose office it is to minister to suffering man, and the experiences collected by each should be willingly laid as common offerings on the altar of science.

To no class of professional men is a liberal education more important than to the army surgeon. To command that respect which is necessary for the right exercise of his official duties, he must be superior in general knowledge to his comrades. The many countries and varied climates to which he is sent, and the delicate positions in which his service often places him, demand the possession of an enlarged and well-stored mind; while the deep responsibility attached to the charge of such a number of valuable lives, and the necessity imposed by the absence of a "consultant" of deciding the most critical cases on his own unaided judgment, demand the firm self-reliance founded on clear knowledge as essential to any measure of success. Even amid the falling ranks, where he is exposed to as great danger as any, he must completely forget self, and give his whole mind to the condition of the sufferers around him; for often do his decisions, formed in a mere instant of time, settle for life or death the fate of the fellow-being before him. Then his powers of observation must be so well trained that he can discriminate between different diseases, whose types are mingled and masked by their union, as these are only seen in armies in the time of war.

The hardships incident to a soldier's life fall equally on the surgeon as upon his comrades: and, besides the dangers of battle and exposure, he runs the risk of those epidemic diseases which devastate armies, and which are the product of exciting causes, to which he has been as liable as any of those actually seized, and to the infection of which, when developed, he is ever exposed. In civil practice, on the other hand, a surgeon is not subjected to those predisposing and exciting causes of disease—cold, want of food and clothing, &c.—which cause its appearance among the mass of the population, nor does he remain exposed to its infection longer than is necessary to prescribe for his patient. The want of libraries for study and self-improvement are also drawbacks to the exercise of the profession in armies, of which the civilian has no experience.

The strict discipline which prevails in military hospitals, gives the army surgeon some advantages over the civilian in the treatment of his cases. No interference from the ill-judged kindness of relatives, or from the headstrong wilfulness of the patient himself, can occur. His opinion is a law from which there is no appeal, and thus fewer obstacles stand in the way of his giving a fair trial to remedies. He has, also, the advantages so often denied the civilian, of correcting or confirming his diagnosis and treatment by after-death examination—a point of the greatest moment. He can, in general, exercise his judgment also to the fullest without having his decision criticised by a host of ignorant censors, and thus the moot points in surgery can often be determined by him in a manner not permissible in civil life.

The greater uniformity in age, constitution, and external circumstances that is to be found among patients in the public services, than among the mass of the population who enter civil hospitals, makes conclusions drawn from their treatment more reliable for future guidance in dealing with them, than any statistics derived from civil practice can be for general purposes.

But how different are the means of treating injury in the field and in civil life! The ample space, established routine, careful nursing, many comforts and appliances of a civil hospital, contrast strongly with the temporary nature, hurried extemporised inventions, and incomplete arrangements of a military hospital in the field.

The influx of patients from the works of a besieging force, or the shifting from place to place of an army during a campaign, makes the removal of the sick to the rear a necessity. Then, as this transference has often to be accomplished by means little adapted for the purpose, and at a period of the treatment the worst fitted for its execution, the evil done is often irreparable; so that injuries which might be completely cured in stationary hospitals have often to be relieved by amputation, while others whose treatment might, under more favourable circumstances, have afforded a fair prospect of success, are placed beyond recovery. From this it follows that the military surgeon cannot always choose either his own time or circumstances in performing his operations. He must be content to do the best he can in the crisis, and thus his experience has sometimes to be sacrificed to expediency. His operations, too, often differ widely from the classic procedures of civil life. The adage, that "a good anatomist may operate in any way," has often in him its illustration. The object being to save as much as possible, compels him to tax his ingenuity in order to take advantage of the eccentric manner in which the ball has half accomplished the severance of the limb, and to seize his flaps here and there where they can be got; and thus, though the immediate result may not appear so satisfactory, the final end is probably as effectively secured. · In the practice of field surgery, moreover, methods of operating will often succeed which are not adapted for civil practice. Thus, in the resection of joints which come to be performed in the field, a comparatively small and simple incision will enable the operator to remove the injured parts, while in those cases in which the operation is commonly performed in civil life, a much larger and more complex incision is generally required in order to permit of the extraction of the enlarged, adherent, unbroken bone which has to be removed, and perhaps to allow of the excision of part of the articular cavity at the same time.

As contrasted with the duties of the naval surgeon, those of the military surgeon are much more difficult. His patients are widely scattered, do not come so soon under his care when injured, are subjected to greater hardships both immediately after being wounded and during treatment, than are the patients of the naval surgeon. "The sailor fights at home," while the unfortunate soldier has often much suffering to go through before he is admitted into hospital.

The soldier as a patient differs from the civilian in several well-marked points. In some respects he is a better patient, and in many respects he is a much worse one. Some of these points of distinction should always be borne in mind when estimating the success of surgery as practised in the case of one or the other.

Chosen when young from the mass of the population on account of his physical promise—selected with care during peace—with less discrimination during war—the soldier at starting is advantageously contrasted with the majority of the men of his own age. Chosen without any reference to his moral character, he is not uncommonly depraved and profligate in his habits, and has perhaps enlisted in the recklessness which succeeds to debauch, or as a last resource to save him from penury. We have thus not unfrequently two conditions meeting in the young recruit, both of which bear their own fruit in his future history—a tendency to indulge in vices which lead to disease, but a state of health in which disease has not been as yet established.

Taken from a domestic life in which he had possibly every liberty as to the disposal of his time, the formation of his habits, and the pursuit of his amusements, he is at once placed under the rigours of a discipline which soon becomes irksome. He enjoys little leisure, but is harrassed by his accustomed, and, for a time at least, laborious duties. Nostalgia succeeds, and thus the period of acclimatization, as it may be termed, becomes an ordeal so trying as in many instances to implant the germs of disease. The prejudicial effects of this initiation will be the more sure, if the recruit be launced into the real business of a war camp before his constitution has had time to accommodate itself to the new condition of things in which it is for the future to exist. But if the young soldier get over this noviciate, then his physical condition, during a time of peace at least, is undoubtedly favourable as contrasted with his fellow in civil life. His food, which is well adapted for his use, is provided for him regularly. He is systematically exercised. His hours of labour and repose are carefully arranged, and he is at all times liberally supplied with fresh air. The civilian, on the other hand, though not subjected to the rough change of existence which the soldier has to undergo, is greatly less regular in his mode of life. He lives frequently in close streets and airless dwellings. His food is irregular, varying with the profits of his labour. He indulges without restraint when he can afford it, and has to submit to privation afterwards to compensate for the excess.

In war, again, the soldier loses many of his advantages over the civilian. The external circumstances which predispose to, or generate disease, are more numerous and vastly more potent in his case than they ever are in civil life. The exposure, the bad and irregular food, the deficient shelter, the excessive fatigue, the unnatural excitement or depression of victory or defeat, all tend to reduce him as much below as he was formerly above the civilian in the scale of health. He has, amidst "the irregularities of war," opportunities for licentiousness of which he is not slow to take advantage, and his unquiet and exciting life is but too apt to occasion that "debility of excess" which conceals a constitution weak to resist injury, under an outward appearance of strength and vigour. Thus it is, that as in civil life different trades produce different diseases, so a soldier's life, both in peace and war, begets its own diseases, and secures exemption from others to which civilians are liable.

Morally, as well as physically, the sick soldier differs from the inmate of a civil hospital. If wounded, he received his injury in the discharge of his duty; if sick, in the fulfilment of praiseworthy service. His "honourable scars" recognize none of those causes referable to misconduct or stupid thoughtlessness, which so frequently make the civilian the inmate of an hospital. He has no fear like the civilian for the future, if incapacitated for further service, as he knows that his misfortune will entitle him to sustenance for the time to come, and that his country will regard him with gratitude.

When struck down by sickness, the soldier is, however, thrown more upon himself than the civilian, and this isolation must in his case act prejudicially on his recovery. He has no visits from sympathizing friends, as he lies on a sick bed, far from home, amidst the selfish hardness of a camp. He is soon separated from his comrades, and placed among strangers gathered like himself from the accidents of the field, and he finds himself in circumstances where he has little to cheer, but much to depress him. In the injuries to which he is exposed in war, he is more hardly dealt with than the civilian. The accidents which befall him, equal in their severity the most terrible which occur in civil life. The effects produced by the massive round shot or ponderous shell, are very like the crushing and tearing of machinery impelled by the resistless steam; so that, among the many assimilating effects of our railways and manufactories, one will evidently be, in course of time, the bringing of the surgery in civil hospitals more and more into conformity with that of war.

But, besides all that I have said as to those matters in which military and civil surgery are similar, or disagree, and as to the contrast which exists on some points between the patients falling to be treated in either case, there are yet some circumstances in the late war to which I must allude, as they are peculiar in themselves, and have an especial bearing on its surgical annals.

A siege differs considerably from ordinary camppaign work, both in the description and mortality of the wounds to which it exposes the soldier. The close proximity of the opposed batteries, the steady and deadly aim which can be obtained by the riflemen, the range so soon ascertained for cannon and mortar, the guns so carefully and accurately worked from the absence of hurry, and from the daily practice of the gunners, all contribute to render the proportion of casualties higher and their severity greater in sieges, than the injuries which attend a campaign in the field. Wounds of the upper half of the body may be expected to be more common in a siege, from the lower parts being protected by the works, and shell wounds must also be of more frequent occurrence, from the larger employment of mortars in attacking or defending a city. The sudden sorties from the beleaguered garrison, the long and constant exposure to the enemy's fire while forming and guarding the trenches, all conduce to swell the number of those injured.

The health of the troops, moreover, does not maintain so high a standard when they are stationary, and want the wholesome animation which results from the change and stirring incidents of a moving campaign; whence it follows, that on becoming inmates of the hospital, they are not so fit to stand active treatment, nor are they so "lively at recovery."

However, there is one advantage which a siege has over a campaign in the field, and it is a considerable one. The hospitals being more stationary can be better arranged, and placed so near the scene of conflict, that the injured may be more quickly succoured.

During the late war, our army had not only to go through the ordeal of great battles, but the prosecution of a siege unequalled for its difficulties in the history of war—a siege in which every obstacle and every trial was enhanced by the stubborn resolution of a brave enemy, and* the frailty of our own military preparation. The sorties were on a scale so gigantic, and pushed so resolutely, as to occasion effects little inferior to those of a pitched battle; and the extraordinary length and active prosecution of the siege caused results resembling those of a constant battle, several months in duration. A few general engagements, and the casualties of outpost service, make up the accidents of an ordinary campaign; but with us, day after day, and night after night, kept up a constant strain, which was more exhausting to the strength of the army than any other sort of warfare could have been.

The majority of the recruits who joined the army early in 1855, and who supplied many of the wounded of that year, were far from being well chosen. They were selected under a pressure, and were the contributions of a country where the drag-net of the conscription is not used to inclose the good as well as the bad, and where a soldier's life is not in any honour or favour with the. generality of tho people. Many of them were raw boys, ill-conditioned, below the standard age, undeveloped in body, unconfirmed in constitution, and hence without stamina or powers of endurance. Often selected on account of their preeocious growth, at once launched into the turmoil, unwonted labour, and hardship of a siege, in which the strength of full-grown men soon failed, they were very quickly "used up." Cholera or fever speedily seized them, overtaxed as they were in work, and unaccustomed to either the food or the exposure which fell to them. The hospitals became filled with such unpromising patients, whose "wizened" look of premature age was remarked by the most casual observer. If these unfortunate boys were severely wounded they almost invariably died, as their weakly constitutions and overstrained powers could not withstand "the ordeal of recovery." To them Hunter's saying applied with peculiar force, that. "their condition of health did not bear disease." If they survived tho first effects of their injury, their convalescence was painfully prolonged, and the least imprudence produced a relapse. Their ailments were seldom acute—their life-power was unequal to its production—their nervous systems were shattered, and that undefined but most fatal disease, known as the "mal des tranches," was soon set up. Depletory measures had soon to be abandoned, and a more rational treatment, founded on special symptoms and the observed effects of remedies, substituted for the conventional medication.

Again, several of the regiments which suffered most in many of the assaults, and which consequently contributed the greater number of the operative cases, were, either wholly or in part, composed of men who had just returned from prolonged service in India. Men so circumstanced were but ill calculated to undergo the rigours of a Crimean winter, or the hard work of the trenches, or yet the great trial of a capital operation.

There was yet another element which demands attention, when estimating the surgical records of the war. I refer to the use of the new rifle, with its conical ball. The rifle used by the Russians was little inferior in range or force to our Minié, while its conical, deep-cupped ball was much heavier. The greater precision in aim, the immensely increased range, the peculiar shape, great force, and unwonted motion imparted by the new rifles to their conical balls, have introduced into the prognosis of gun-shot wounds an element of the utmost importance. I am not prepared to say whether the great destruction of the soft and hard tissues which these balls occasion, results from their wedgelike shape, immense force and velocity, or the revolving motion, or from a combination of all these causes combined; but of one thing I am convinced, that their use has changed the bearing of many points which fall to be considered by the surgeon in the field. The severity of the primary action on the part struck, and especially the aggravated evils which follow their wounds, combined to exercise a most prejudicial influence on the surgery of the war, to which due weight has never been given. Immense comminution of bone has been their most prominent effect. The amount of laceration of the soft parts seems to depend on the distance at which the missile is fired.

The wide-spread destruction of the bone often renders consolidation impossible, so that amputation has more frequently to be had recourse to, and the distance from the trunk at which that operation has to be performed being diminished by the same causes, the resulting mortality has been greatly increased. All who compared the dead of this with former wars, especially of Indian battles, were painfully struck with the greater disfigurement of the corpse caused by the conical, than by any other species of ball.

But besides the more destructive nature of the small-arms employed, cannons and mortars were used on both sides, of a calibre and range never before tried in any war. When Parè thought the cannon of his day so enormous and destructive, what can we say of those huge sea-service mortars and immense cannon used to defend and attack Sebastopol, compared with which those of the last century are as toys!* The fragments of our modern shells must be as weighty as the whole projectile known to our forefathers, and the grape which was so freely used in the East were half as large as the round shot fired from the field guns in the Peninsula. With us, every refinement in the art of destruction was liberally practised, so that "l'art de tuer les hommes avec methode, et gloire," was, unhappily, never carried nearer perfection, though we may comfort ourselves with the reflection of Percy, that this very perfection, "nous a donné la même tâche et la même recompense dans l'art de les conserver." "Les circonstances," says Briot, "qui contribuent le plus à la destruction des hommes sont aussi celles qui font decouvrir et developpent plus de moyens propres à leur conservation."

Finally, if in war the surgeon sees much which is terrible, much which taxes his feelings of humanity, and his regret at the feebleness of his art, he has also the comforting conviction that

---

* "Truly," says Parè, "when I speak of the machines which the ancients used for assaulting men in combats and encounters, it appears to me as if I spoke of infants' toys in comparison with these, which, to speak literally, surpass in figure and cruelty the things which they thought the most cruel."

nowhere is his beneficent mission so felt, nowhere is the saving power of his profession so fully exercised; so true is it, that " Chirurgery triumphs in armies and in sieges. 'Tis there that its empire is owned, 'tis there that its effects, and not words, express its eulogium." *

---

* Dionis, quoted by Sir George Ballingall.

CHAPTER II.

The "peculiarities" of Gun-shot Wounds, and their general Treatment.

In saying that "there is a peculiarity, but no mystery, in gun-shot wounds," John Bell has expressed the change of opinion which late times have brought about with regard to the nature of these injuries. It was the mysterious character ascribed by the old surgeons to wounds from so "devilish an engine" as a gun, which so long surrounded them with dread, and made incantations and charms the favourite resource in their treatment. The new philosophy has dispelled the mystery, but left us still to study the eccentricities which so often mark these injuries. The contused appearance and unavoidable sloughing of the walls of the ball's track, the little-suspected, but serious destruction of deep parts, and the grave consequences which may ensue from such a wound, appear to have been the circumstances that suggested the envenomed nature of gunpowder, and the cautery-like action of its projected ball, as well as the idea which prevailed, that in order to get quit of the injurious influences thus exerted on the wound, it was necessary to pour into it burning oil, or curious tinctures concocted from the most opposite and absurd ingredients, or to smear the part with nauseous grease and "charmed salves."

The description of the sensation caused by a gun-shot wound in a fleshy part, usually given by the sufferer, is, that it resembles the effect of a smart blow from a supple cane. Some, however, feel as if a red-hot wire were passed through the part. The fracturing or splintering of a bone is always more painful than a flesh wound, and if a joint or larger cavity be penetrated, the pain is still more acute, and the shock still greater—in most cases proportioned to the vitality of the part injured.

It is a very remarkable, though universally known fact, that when the mind is greatly engrossed by external objects—excited "'mid the current of the heady fight," severe wounds may be received without any consciousness on the part of the receiver. Whether the sensation may be so very slight as to be immediately obliterated by the tide of strong passions rushing through the mind of the combatant, or whether a reflex act of the mind be necessary for receiving a sensation—in common words, for perceiving the state of its companion, the body—I shall not attempt to discuss. But all military surgeons will confirm the statement of Hennen, when he says that "some men will have a limb carried off, or shattered to pieces by a cannon-ball, without exhibiting the slightest signs of mental or corporeal agitation—nay, without being *conscious* of it." I myself have known an officer who had both legs carried away, and who said that it was only when he attempted to rise, he became aware of the injury he had received; and very many who had suffered slighter wounds, have said that the trickling of blood along the skin was what first called their attention to their state.

The "collapse" and the mental trepidation which frequently follow the infliction of a mortal wound in the trunk, are, in many cases, most appalling. But although the presence or absence of this severe constitutional effect is useful as a diagnostic indication of the gravity of the injury, it is not entirely to be depended on, for the terror and amount of "shock" frequently depend as much on the "nerve" and frame of the sufferer, as on the severity of the wound. The different effects produced on different persons by wounds in every respect alike, are obvious to every one who has seen war, and call for the exercise of a most discriminating judgment on the part of the surgeon. Then, the period of "collapse," which will, to some degree, occur in every case of a severe wound, varies greatly, which must determine whether immediate amputation be necessary, or whether it would be safe to delay it. The only other remark we make on this subject is, that the "commotion" succeeding gun-shot wounds is greater when the lower extremities are injured, than when the arms suffer; and this is more especially seen, if the person be in an erect position when the injury is inflicted; which observation is consistent with the remark made by Chevalier, that the shock is always greater when the ball strikes a muscle in action, than when it impinges against one which is relaxed.

The destruction inflicted by a ball depends on the distance at which it is fired, the direction of its flight, its shape and velocity, as well as on the nature of the part struck. If fragments of metal are fired, as sometimes happened during the sieges of the Peninsula, as well as in the civil emeutes of Paris, and of which we had some experience in the Crimea also, a very lacerated, irregular, and dangerous wound may be caused. A ball passing at great speed over the surface of a limb, may occasion a wound similar to that made by a knife. But this action of a ball is rare.

The great velocity, peculiar shape, and motion of the conical ball, give to its wounds a character considerably different from those which is present in wounds caused by a round musket ball. If fired at short range, and if it strike a fleshy part, the conical ball produces, I think, less laceration of the soft parts than the old ball; but if the range be great, and the part struck bony, with little covering of flesh, as in the case of the hand or foot, then the tearing, especially at the place of exit, is greatly more marked.

I have not been able to satisfy myself in all cases, so clearly as the description of authors would lead me to suppose I could, as to the characteristics which distinguish the wound of entrance from that of exit. That the former is more regular and less discoloured than the latter, is true in many cases, but that the lips of one wound are inverted, while those of the other are everted, has seldom been clearly marked to my observation. If the speed of the ball be great, and no bone have been struck, then there is little difference in either the size or discoloration of the wounds; but if the flight of the projectile be so far spent as to be retarded by contact with the body, especially if it have encountered a bone or a strong aponeurosis, so that its speed is considerably diminished before it passes out of the body, then the wound of exit will considerably exceed in size that of entrance. This is especially true of conical balls. If, on the contrary, the ball be fired close at hand, so that its speed is not sensibly diminished by its passage through a limb, then the difference of size will be very small, and may even be in favour of the wound of entrance, as I had twice an opportunity of observing.

The usual action of a ball in proportioning the size of the two orifices is easily understood, when we consider that the part of entrance is supported by the whole thickness of the limb, while that of escape is quite unsustained, and therefore the more liable to be torn. Huguier has shown that the loss of substance which occurs at the place of entrance, and the flap-like tearing which takes place at the orifice of exit, form the best marks of recognition we possess, and that these characters can always be made out by examination of the clothes or accoutrements traversed in cases in which the supervention of inflammation has effaced them from the wound itself. The introduction, but non-escape of a foreign body, as a piece of the breastplate, belt, buckle, or part of the musket, &c., along with the ball, which alone passes out; or the flattening of the ball against a bone within, and its diameter being thus increased before it escapes, will all contribute to vary the relative characters of the orifices of the wound.*

To the military surgeon, it is often of consequence to be able to conclude whether the two apertures in his patient's limb have been occasioned by one ball, which is thus seen to have passed

---

*In Arnel's experiments given in the *Journal Univer. de Med.* for 1830, it is shown that a ball, fired against a number of planks firmly bound together, causes a series of holes progressively increasing in size, so that a cone is formed by their union, whose base is represented by the last exit hole. M. Devergie's experiments on the same point, given in his communication to the Academy, go to prove this also. Velpeau and others have objected, but without good grounds, to the deductions drawn from the experiments being applied to the question.

out, or by two balls still imbedded in the limb, and to the medico-legal jurist, the knowledge of the marks which characterize the two wounds, is of much moment.*

The action of a ball on the different tissues of the body may be, in a great measure, inferred from a consideration of the shape of the projectile, and the nature of the part struck. It carries away, as I before remarked, a piece of the skin at the place of entrance, and rends it where it escapes. The small plug of integument which is carried into the wound, Huguier tells us can often be discovered there.†

The contusion which a ball causes in traversing *muscle* gives rise to one marked characteristic of gun-shot wounds—their healing only by suppuration and granulation. Occasionally an exception occurs to this rule. Thus, I have seen a case in which a superficial wound of the gastrocnemius was said to have healed without suppuration by the fifth day, and in the records of a Sepoy regiment in India, I find mention of even a deeper gun-shot wound of the deltoid healing in the same way by first intention.

Dr. Stewart, staff assistant surgeon, reports‡ a case of a similar union, as having been observed by him during the Caffre war. A Fingo received a pretty severe gun-shot wound of the muscles of the back, and union without suppuration took place. Two things are necessary to produce such a happy result: 1st, a most healthy and temperate patient; and 2nd, the rapid flight of the ball.

It is curious to notice how large a body may enter through a muscle, and hide itself without producing any great wound. Thus, I saw a case at Scutari, in which a piece of shell, weighing nearly three pounds, was extracted from the hip of a man

---

* Between the opposite views held by Blendin and Dupuytren, the opinions of military surgeons and medical jurists have oscillated, evidently from the fact that no constant relations exist between the entrance and exit wounds. Velpeau, holding a middle view, concludes with truth, "Dupuytren is wrong, and his antagonist is not right." The distance at which the gun is fired has most to do in determining their character, according to Devergie, who has himself, however, recorded a case which proves that the wound of entrance may be the larger, even when the gun is fired at a distance. Begin has given us the following valuable observation, with regard to the resulting cicatrices. That of entrance, he says, is generally white, depressed, and often adherent to the underlying parts, while that of exit is only a sort of irregular spot which does not adhere to the parts below, and is sometimes so indistinct as to be concealed in the folds of the skin. This difference he explains by the loss of substance sustained at the point of entrance.

† John Hunter speaks also of this piece of detached integument.
‡ Unpublished records of medical department.

wounded at the Alma, which had been overlooked for a couple of months, and to which but a small opening led. Larrey gives a case in which a ball, weighing *five* pounds, was extracted by him from the thigh of a soldier. The presence of so large a body had not been detected by the surgeon in charge, and the patient suffered no inconvenience from it beyond a feeling of weight in the limb. Paillard mentions, having heard M. Begin recount a case in which a ball of *nine* pounds so buried itself for a time. Hennen, too, mentions a case as having occurred at Seringapatam, in which a spent *twelve* pound shot, buried itself in the thigh of an officer, and "so little appearance was there of a body of such bulk, that he was brought to the camp, where he soon expired, without any suspicion of the presence of the ball till it was discovered on examination." It is more easy to understand how a large fragment of shell should so conceal itself than a round shot, as if its long diameter corresponded with the run of the fibres of the superficial muscles, and especially if the muscle was relaxed at the time of contact, then a large piece might enter a muscular limb without causing an amount of injury proportioned to the size of the body introduced. Such an instance occurred in the Crimea to a French soldier, of whose case Baudens has given an account. A fragment of shell, weighing 2 kilog. 150 grammes, so completely buried itself in the thigh as almost to be invisible. The elasticity of the soft parts doubtless assist in closing the opening by which such a mass entered.

Baudens has made an observation which I am not aware has been confirmed by any other, viz.: that when the ball is cut out from among the muscles, however early it may be accomplished, it has a cellular envelope round it, which he calls "kyste primatif," as contrasted with the "kyste definitif," which forms its sac when it has been long inclosed in the tissues.

Muscles which have been severely injured by ball are very apt to become contracted during cure, if precautions are not taken to prevent it. Of this most disagreeable result I have seen a good many cases in the East.

On *tendons* a ball may cause little or no injury, especially if they be relaxed at the moment they are struck. Their toughness, elasticity, form, and mobility, all help in protecting them from being cut across, or pierced. A round ball is often deflected by a strong aponeurosis like the "fascia lata," particularly if it strike at an angle to the surface, and at a period of its flight when the force is somewhat expended. A conical ball, however, is seldom so turned.

It is on *bone* that the destructive effects of a ball become most evident. (1.) When its line of flight is very oblique, and it is a flat bone against which it strikes, then it may be thrown off, causing no other damage than depriving the bone of its periosteum. When this occurs in the case of bones of the head, much danger may subsequently ensue, as will afterwards be shown.

Contused wounds of the long bones, though seemingly of little moment at first, are sometimes very serious in their results, not only from the separation of the periosteum, and subsequent disease of the bone arising from that source, but also from inflammation being set up in the medullary canal. (2.) A round ball may be flattened against the shaft of a long bone, without causing any subsequent harm. This was often seen in India, where the matchlock is used. (3.) It may turn round a bone without breaking it. Thus, Chevalier records a case in which a ball, entering at the lower part of the thigh, passed spirally round the bone to the top of the limb, "comprehending nearly the whole length of the bone in one circumvolution." (4.) A round ball, as is well known, may notch or partly perforate a long bone without causing fracture, and pass off, or remain in the medullary cavity, having passed through the outer wall. This is, as can be easily understood, a most dangerous accident. (5.) If the force of propulsion be a little greater, then the bone may be split longitudinally, without being fractured across, as in a case related by Leveillé, and quoted by Malgaigne, in which an Austrian soldier at Marengo was struck by a ball in the lower third of the leg. He walked several miles to the rear, where he was seen, and the wound thought to be very slight. A superficial exfoliation of the bone was alone expected; however, his symptoms became so serious that the leg had to be removed, when it was found that, from the place where "the impression of the ball" existed, there proceeded several longitudinal and oblique clefts, which extended from the lower third of the tibia up to near the head of the bone. (6.) Into the spongy heads of bones, and, more rarely, into their shafts, a ball may be driven as into a plank of wood, without almost any splintering, and become encysted there. (7.) It may pass through, causing a clean hole, of several of which occurrences I will afterwards relate cases; but the conical ball never acts in any of these ways, so far as I have seen. It is seldom split itself, but invariably splinters the bone against which it strikes to a greater or less degree, according to circumstances, and that in the direction of the bone's axis. This tendency to splitting in the bone shows itself much more in a downward than in an upward direction, so that the destruction which such a ball will occasion will be greater when it strikes the upper than the lower end of a shaft.

All kinds of balls generally fracture and split the shaft of a bone if they strike it about its middle, but while a fracture with but little comminution results from the round ball, the conical ball—especially that which has a broad deep cup in its base—splits and rends the bone so extensively, that narrow fragments, many inches in length, are detached, and lesser portions are thrown in all directions, crosswise at the seat of fracture, and driven into the neighbouring soft parts. It is the pressure of these fragments, as will be shown further on, which renders the fracture of long bones by the new ball so hopeless.* I had many most interesting opportunities of seeing the extraordinary manner in which the conical ball destroys bone in the way I refer to. I have never met with an instance in which such a ball, fired at whatever range, and striking at all perpendicularly on a long bone, has failed to traverse it and comminute it extensively.

From the comparatively little employment of the round ball during the late war, there were fewer illustrations of the splitting of balls on the edge of bone, as, for instance, on the edge of the tibia, or on the bridge of the nose, or on the humerus, than usually occur in a campaign. I do not believe that the conical ball, with its immense force of propulsion, could be so split. There is a case borrowed from Mr. Wall of the 38th, given later under wounds of the head, in which "a round rifle (?) ball" was thus split on the parietal bone, one-half entering and the other half going off externally, in a soldier of the 38th, wounded on the 8th September. Another somewhat similar case occurred in the 19th regiment. It is by no means uncommon that a ball should be thus split on the head. Many examples of it occur in-works on military surgery. No case clearly made out as one of splitting came under my own notice; but in one instance, a ball so changed in shape as to appear the section of one, was extracted from within the iliac fossa. Instances are on record in which balls have been split into three parts by the bones of the face, and the trochanter major.

Although it cannot be for a moment doubted that balls may remain for a lifetime imbedded in bone, and cause little, if any annoyance, yet it is equally certain that the most grievous results much more frequently arise from their presence in such situations. Of this, innumerable examples readily occur to any one who has seen many "veterans;" or who has read much on

---

* As instances of how great a difference it makes in the prognosis of cases whether a round or a conical ball has been the wounding agent, I may relate two cases from a host of others. In the first instance, the ball entered on the external side of the ankle, near the tendo-achillis, and passing forwards and inwards, lodged, as if in a piece of wood, in the lower end of the tibia, close over the ankle joint. When the ball was removed, the bone was found not to have been split in any direction. A conical ball would have, to a certainty, opened the joint, and, in all probability, so split the tibia as to have necessitated amputation in the upper part of the leg. In another case, a round ball made a clean hole through one of the condyles of the femur, and did not split the bone; while, if a conical ball had struck the same part, it would have so cleft the bone that amputation in the middle of the femur would have been called for; whereas, the removal of the limb at the knee joint—a much less serious operation—sufficed in the case referred to.

the subject to which I refer. When speaking of wounds of the shoulder joint, I will detail some cases which illustrate the pernicious action of balls left impacted in bone. Guthrie is very emphatic in his directions to remove balls so placed, and predicts the most disastrous consequences from the neglect of this measure. Malgaigne, after relating several cases in which balls have remained without causing harm, concludes thus: "It is necessary to mention these fortunate cases as evidence of the resources of nature, but they hardly serve to weaken the force of the prognosis when a ball cannot be extracted, or the essential indication of this sort of lesion—the extraction of the foreign body. This indication is, then, that of the first importance."

The *nerves* most commonly escape injury from a ball. If the missile has been rendered irregular in shape by previous contact with some hard substance, then it may do much damage to even the larger nerve trunks. Numbness, succeeded by pain in the extremity of a limb traversed by a ball is not uncommon, and probably arises from the contusion or laceration of some chief nerve—the swelling and the pressure it occasions assisting to give rise to the subsequent uneasiness. The paralysis which succeeds the injury of a nerve may come on at once, or after an interval, and may, or may not, be accompanied with pain in the part itself, or in other regions connected with it by nervous communication. I have seen the hand several times waste when some of its nerves had been injured by a ball. In one case in particular, in which the ball had coursed up under the muscles on the external surface of the upper arm, this symptom was very marked.

Even though making all due allowance for the elasticity, strong coat, mobility, and form of the *arteries*, it is yet difficult to understand how they escape injury in gun-shot wounds as they do. The rarity of primary hœmorrhage on the field of battle has been long remarked, and yet how often do we meet with ball wounds apparently through the course of a great vessel!

The veins are more easily cut than the arteries, and primary hœmorrhage, when it does occur, proceeds more commonly from them. Some vessels are more liable to injury from balls than others. Thus, those firmly tied down, or lying on bone, are more subject to damage than those loosely reposing on the soft tissues. This remark applies especially to two vessels, the femoral as it passes over the brim of the pelvis, and the popliteal, where it lies on the head of the tibia. The lower parts of the ulnar, the radial, and the facial, where it turns over the jaw, are subject to injury from the same reason. An artery has not rarely been opened by a spiculum of bone detached by a ball which had itself spared the artery.

The *eccentric course* often pursued by balls has been a frequent subject of remark, and though we had many most striking instances of this, still I suspect we have had less of it than occurred in the experience of former wars. The conical ball seldom fails to take the shortest cut through a cavity or limb, and it has at times been seen (as at the Alma) to pass through the bodies of two men and lodge in that of the third. But of the wanderings of the old round ball there were many illustrations. I have known it enter above the elbow, and be removed from the opposite axilla; and in another case it entered the right hip, and was found in the left popliteal space.* This "bizarrerie" in a ball's course is accounted for by the deflecting action of tendons, aponeuroses, or processes of bone, or by the angle at which the ball strikes, and the way in which, during certain positions of the body, distant parts are placed in a line, as in the well-known case recorded by Hennen, in which a ball entering the upper arm of a man ascending a scaling ladder, was found half way down the thigh of the opposite side. The fact of this wandering, however, is a peculiarity in gun-shot wounds which often renders the discovery of the wounding agent difficult.

Foreign bodies, as pieces of cloth or part of the soldier's accoutrements, are often far more troublesome when introduced into a wound than the ball which occasioned their presence there. Innumerable and most heterogeneous have been the foreign bodies thus forced into wounds; but those which are capable of acting chemically as well as mechanically, are the worst of all. Of these, lime, pieces of copper, &c., are the most frequently met with. Round lead balls are, perhaps, from their nature and shape, the least noxious of any, and are most likely to become encysted in the tissues.

Few questions connected with gun-shot wounds have given rise to so much discussion and diversity of opinion as that with reference to the *extraction of balls*. For my own part, I have seen enough to make me subscribe, with all sincerity, to Begin's precept, when he says in his communication to the Academy: "Selon moi l'indication de leur extraction est toujours presente, toujours le chirugeon doit chercher a la remplir; mais il doit lo faire avec la prudence et la measure que la raison conseille. S'il recussit, il aura beaucoup fait en faveur du blesse. S'il s'arrête devant l'impossibilite absolue ou devant la crainte de produire les

---

* The surgeon of the 24th, when serving in India, mentions a case in one of his reports, in which a ball entered below the angle of the lower jaw, on the left side and made its exit above the spine of the right scapula, without injuring any important part; and M. Meniere, in his account of the Hotel Dieu during the "three days," tells us of a ball which entered at the inner angle of the left eye, passed downwards, backwards, and to the right side, under the base of the cranium, and was removed above the right shoulder. The rapid recovery, without a bad symptom, was no less wonderful in this case than the direction taken by the missile was curious.

lesions additionelles trop graves il aura encore satisfait aux principes de l'art; et quels que soient les resultats de la blessure il n'aura pas a se reprocher de les avoir laisse devenir funestes par son inertie."

If we examine into the opinions of surgeons on this point, we find that nearly all those who look upon the extraction of the ball as a matter of secondary importance are civilians, while military surgeons place great weight upon its accomplishment. The true way of putting the question is, not whether balls may remain in the body without causing annoyance, but whether they do so in so large a number of cases as to warrant non-interference. We must always remember that "science is not made up of exceptions," but is established by a collection of positive facts. Those who have studied gun-shot wounds in the field, know full well how enormous is the irritability caused by the presence in a wound of a ball or other foreign body—how restless and irritable the patient is till it is removed—how prolonged the period of treatment is in the cases in which it is left—and how frequently the results are so distressing as to demand future interference, or condemn the unfortunate sufferer to a life of discomfort. As it is the surgeon's duty to treat his patients with reference to their future case as well as to their present cure, so he should not try to bring about a healing of the wound which can be only temporary and fallacious, to the sacrifice of the efficiency of a limb and the future health of the body.

In this country we have not many opportunities of obtaining extensive information on the point as connected with the subsequent history of men with balls remaining unextracted, but such information is supplied from the Hotel des Invalides of France, by M. Hutin, the chief surgeon to that magnificent establishment. He tells us, that while 4,000 cases had been examined by him in five years, only twelve men presented themselves who suffered no inconvenience from unextracted balls, and the wounds of 200 continued to open and close continually till the foreign body had been removed. This epitome is of much value in estimating the question I am considering. In leaving the ball unextracted, we never know what evils may follow. The keeping open of the wound exposes the patients in the first place to all the dangers of a life in hospital, and the very elimination of the foreign body by suppuration, if it take place at all, necessitates a vast amount of annoyance. If it be a piece of shell or such like which is present, then its size will prevent its unaided extrusion, and the blocking up of the track which it is so apt to occasion, may cause burrowing abscesses of a most destructive character.

Before a ball becomes encysted, it may set up grave inflammation, which will mat together and embarrass parts; press upon bone, and perhaps cause exfoliation; ulcerate blood vessels, and so irritate nerves as to occasion affections as severe and fatal in their results as tetanus. It is somewhat remarkable, that in the wounded who came under my own care, two died of tetanus, in the very small number of instances—four or five at most—in which I could not find the ball. If this was a mere coincidence, it is the more curious. Gravitation and muscular action may so change the position of a ball, that from a harmless site it may be removed to one of much danger. It may thus work its way into a cavity, and cause fatal results.

But suppose the ball to become encysted in the first instance, what security have we that some very trivial circumstance (it may be a blow or even a deterioration in the health of the patient,) may not set up irritation, inflammation and suppuration in the cyst, and so come to set the ball free again to work harm in the economy? In any case, its continued pressure gives rise to much uneasiness. The constant weight and weakness felt in the limb, the wandering pains, ascribed to rheumatism from their aggravation by cold and damp, which attack even distant parts of the extremity, and the ever-present dread felt by the patient, if the ball be in close neighborhood to any vital organ, all unite to give much annoyance and discomfort.

The aversion which patients who have long carried unextracted balls express to have them removed, is not, as some would try to show, any proof of the slight annoyance they occasion, but simply indicates that they choose to suffer the discomfort rather than undergo what appears to them an uncertain and dangerous proceeding to free themselves of a bearable inconvenience.

It seems, then, the teaching of experience, as it is of common sense, that whether the question be viewed as one bearing immediately or remotely on the result—on the *cure* of the patient, in the proper acceptation of the term—then we should, as soon as practicable, ascertain the position of the ball, remove it along with any other foreign body which may have been introduced with it, always supposing that by such a proceeding we do not cause more serious mischief than experience shows the presence and after effects of the ball can produce.

To extract a ball is, in general, not difficult. It is of much consequence to proceed to its accomplishment before inflammation and swelling have come on, so as to close the wound* The great point to attend to undoubtedly is the fulfilment of the rule, which is as old as Hippocrates, to place the patient as nearly as possible in the same position as that he occupied at the moment of injury—to put the same muscles into action, and the angle which the parts form to one another in the same relation; also, to place ourselves relatively to him in a position to corres-

---

* Percy adds another reason to encourage us in the early removal of balls, when he says that men submit the more readily soon after the receipt of the wound to the necessary incisions, before their courage has been broken by pain and suppuration.

pond as nearly as possible with the direction from which the ball came. By considering the effect which bones or strong tendinous expansions may have had in deflecting the ball, or by paying attention to what Guthrie calls the general "anatomy of the whole circle of injury," and consulting the patient's own ideas, which often convey to us most useful hints, we shall in general succeed without much difficulty in discovering the ball. An examination of the patient's clothes will show us whether any part of them has been carried into and left in the wound—whether the two holes seen in the limb have been caused by the same ball which has thus passed out, or by two balls which are still in; as well as whether the ball may not have carried in a *cul-de-sac* of the clothes, and been withdrawn with it. If this be not attended to, very awkward mistakes may be made, as the mere correspondence in the direction of the two apertures, any more than their seeming want of relationship, cannot be taken as decisive in settling the matter. This point is well illustrated in the following instance, related by an Indian surgeon: A wound was found below, and another above, the patella of a wounded man. The former had all the signs of the wound of entrance, and the latter those usually found at the place of exit of a ball. The opening of an abscess, which formed in the thigh a fortnight after, gave exit to a grapeshot, and it was found that the external condyle had been injured, and that each opening had been caused by a different ball.

In another instance, which occurred in the case of a soldier of the 40th regiment in Cabul,* the ball appeared to have passed through the elbow joint, and to have fractured the radius. There were two openings, having all the appearance of being those of entrance and exit; yet the ball was found and removed from the limb three weeks after. Such a mistake is most apt to arise when two balls have been fired together from the same gun, which happens not uncommonly in civil commotions, or when such fire-arms are used as the "espignole" of the Danes, from which a number of balls are fired in rapid succession, or when a cartridge, similar to that used during the Schleswic-Holstein war, is employed, in which two balls and a piece of lead are bound up together. One ball, too, it should be remembered, may make several openings. Thus, I have seen two in the leg and two in the hip, and also two in either thigh, occasioned in each case by one ball. Dupuytren relates a case in which, from its splitting, one ball made five holes; and the younger Larry saw at Antwerp six orifices caused in the same way. Sir Stephen Hammick mentions a case in which an aperture was found on either side of the chest of an officer shot in a duel. These corresponded both in position and character to those which would

* Unpublished Report.

be occasioned by a ball that had traversed the chest; yet, after death, two balls were found in the body.

As showing the necessity of an early and careful search, as well as that we should never rely too much on the patient's statement, I may mention the following case: A soldier, wounded on the 18th June, came under my care in the general hospital. His right arm, which had been fractured compoundly, was greatly swollen at the time of admission. I was told, and accepted the story, that the accident had been caused by a piece of shell, to which species of injury the wound bore every resemblance, and that it had been removed by a surgeon in one of the trenches. At the earnest solicitation of the patient, I contented myself with applying the apparatus necessary to save the limb without minutely examining the wound. The injury turned out to be much masked, and to be greatly more severe than it at first appeared, the shaft of the humerus having been split into the joint. When removing the limb at the shoulder, some days after, to my great astonishment a large grape shot dropped from among the muscles.* I before alluded to another case in which a piece of shell, weighing nearly three pounds, had remained concealed for two months without suspicion, from a like neglect of a proper examination.

It is well to remember, also, in searching for balls, that they may have dropped out by the same aperture by which they entered, before we come to examine the case. Stromeyer has put us upon our guard against very curious errors, which he says he has seen made in cutting on the head of the fibula, and on a metatarsal bone for balls.

Sir Charles Bell has shown how the nerves may indicate to us the position of the ball. In one case he found it by pressing on the radial nerve, and so discovered that the ball lay behind it. "So when a ball has taken its course through the pelvis or across the shoulder, the defect of feeling in the extremity, being studied anatomically, will inform you of its course—that it has cut or is pressing on a certain trunk of nerve."

From all this, then, it is at the least very evident that we should not be too hasty in concluding that no ball remains in the limb, even although all the signs usually indicative of its having escaped are present; and also, that immediately before proceeding to take any steps for the removal of a ball, we should make certain of its position, remembering the rule laid down by Dupuytren—never to act upon information regarding the site of a ball obtained the day before, from the rapid manner in which they often shift from one spot to another.

---

* I may, however, remark that this splitting upwards of the head from the shaft is very rare. In general, the splitting ceases at the epiphysis.

The common dressing forceps, if long enough and fine enough in the handle, will, I believe, be found the most useful bullet extractor. That invented by Mr. Tuffnel, of Dublin, acted well in the few cases in which I tried it. Larrey employed polypus forceps in preference to anything else, but the inventions which have been made to accomplish this simple end are innumerable. To support the limb with the disengaged hand on the side opposite to that at which we introduce the forceps, is of much importance. If the course of the ball has been from above downwards, and if it has approached at all near the surface, it should always be cut upon the dependent part, by which two objects are secured—the removal is facilitated, and an opening for the pus is insured. If the wound be large, as it generally is from the conical ball, the finger forms the best probe, both to discover the ball, and also to examine the state of the adjoining parts; otherwise, a large gum elastic boughie is our best resource. Causing the patient to move his limb, sometimes makes the site of the ball be felt by him, if not by us. Its position under a fascia, or in contact with a bone, would make us risk much in order to remove it.

The contentment of mind which results from the extraction assists recovery amazingly. The long continuance of the discharge, its gleety character, and the persistence of pain in the track, almost always proceed from the presence of some foreign body—it may be a mere shred—in the wound. Chloroform is of inestimable service to us, both in the examination of wounds, and in the removal of balls. All those voluntary muscular contractions which, although they are apt to interpose obstacles to our examination, were not presented to the entering ball, are done away with, and the severe pain which a prolonged examination and difficult extraction give rise to, is avoided. We must, however, be careful to obtain from the patient all the information he can give us, before we bring him under the influence of the anæsthetic.

The inflammation which ensues in a gun-shot wound shortly after its infliction, makes itself visible in the swelling and consequent eversion of the lips of both entrance and exit wounds, in the general tumefaction of the parts, and in the augmented pain. It was the fear of this inflammation strangulating the parts which gave rise to the exploded custom of scarifying the wound.* The swelling will differ much in different regions and in different constitutions. In parts strongly bound down, in irritable tissues,

---

* Hunter expresses, with his usual clearness, the principles which should guide us in enlarging a wound, or "scarifying," as it was called. "No wound," he says, "let it be ever so small, should be made larger, except when preparatory to something else, which will imply a complicated wound, and which is to be treated accordingly."

in lax distensible parts it will vary much, while, according as the patient is of an inflammatory, lymphatic, or nervous temperament, the effect will differ not a little.

The constitutional fever which sets in is generally proportioned to the importance of the part implicated, though most anomalous exceptions do occur. This fever will often put on the characters of the endemic or epidemic fever; but in war the tendency seems generally to be to a low typhoid type, unless there be a decided local influence in action, as that arising from paludal emanations. With us, the symptomatic fever must have been comparatively slight and evanescent to what it was in the Peninsula. The severe remedies put in force by the surgeons of Wellington's army never could have been employed by us. That old soldiers, if sober, are much less affected by this constitutional disturbance than others, is, I think, very observable.

The mitigation of the constitutional fever, and of the local inflammation, the prevention of all accumulations of matter by making judicious escapes for it, the relaxation of severed muscular fibre, the application of light unirritating dressings, rest, and attention to the essential principles of all surgery, comprise the general treatment which gun-shot wounds usually demand. In the early stages cold may be of use locally—even ice, as recommended by Baudens—and in wounds of the hand and forearm irrigation is of the greatest service; but when inflammation and suppuration are present, hot applications will always be found of most good. Strict attention to the position of the limb is of great consequence, and though in general it may be desirable, as in some instances it is absolutely necessary, to restrict the diet, yet in those cases in which much suppuration is to be expected, very great latitude should be observed with reference to such a rule. Soldiers in war are commonly easily depressed, and should not be fed too sparingly when admitted into hospital, unless they suffer from a wound of the head, chest, or abdomen. Without placing too much faith on the happy effects which Malgaigne tells us the Russian wounded, treated in Paris in 1814, derived from a stimulant diet, as contrasted with the Prussians, French and Austrians, still it is unquestionable that there is too much tendency to look on the common run of gun-shot wounds as highly inflammatory, and to treat them accordingly. Velpeau's rule on this point agrees with his usual intelligent views, when he says he lays it down as a rule to remove his wounded and operated on as little as possible from their ordinary diet

---

It should not be opened because it is a wound, but because there is something necessary to be done which cannot be executed unless the wound is enlarged. This is common surgery, and ought also to be military surgery respecting gun-shot wounds."—*Hunter's Collected Works*, vol. iii., p. 549.

when they are hungry, and when there is no disturbance of the digestive and circulatory systems.

We found codliver oil of the greatest use in those cases in which the waste by discharge was great. A stream of lukewarm water made to pass by gentle syringing from one opening to the other, forms one of the most useful methods of treating gun-shot wounds. Any shreds of cloth, clots of blood, pus, &c., which may be in the wound and sustain the suppuration, will thus be got rid of with very little disturbance to the parts. The addition of a little of Burnett's solution to the water thus used, or to the water-dressing, was useful at the same stage. Of the tonic and stimulant injections recommended by writers I had no experience; but I have seen the French employ, with apparent advantage, a lotion composed of one part of perchloride of iron and three of water in profusely suppurating wounds.

The extreme simplicity of the appliances and dressings employed during this war, and the nearly total absence of poultices, and such like "cover sluts," would, I think, have pleased Mr. Guthrie. The "stuffing in of great tents" was, I need not say, unknown; and though we ascribed wondrous virtues to cold water, it was not on account of any "magical or unchristian" power which we supposed it to possess. Water-dressings, and the lightest possible bandaging consistent with the fulfilment of well-understood ends, were prevalent in our army, but not to the same extent among our allies, who have not yet given up the weighty pledgets of Charpe, and much fine linen, which so greatly astonish the English surgeon. The splints and other apparatus used partook of the simplicity of the rest of the treatment. Stiff bandages were too little used, if we accept the experience of the Schleswic-Holstein war; but the difficulty of always getting the necessary materials in the field is somewhat opposed to their use.

The state of the weather has got much to do with the rapid cure of gun-shot, as of all other wounds. From a perusal of the medical records of regiments serving in the colonies, it would, however, appear that hot weather, as in India, is, on the whole, more favorable than a cold climate.

Shell wounds, and grazes by round shot, are often followed by much injury, little suspected, but deeply seated, resulting, not unfrequently, in wide-spread sloughing of the soft parts. I cannot avoid relating the following case, although it did not occur in the Crimea, as it is a most excellent illustration not only of the great and, it may be, little suspected harm which may be occasioned by a round shot, but also because it is an instance of what would have been in former times set down to the wind of the ball. It is from the records of the medical department. Private John Conally was hit at Sadoolapore by a round shot on the outer side of the right arm and thorax. A blue mark alone was occasioned on the arm, and little or no mark was found on the chest. He died in twenty hours, without having rallied from the shock. The peritoneal cavity was found full of dark blood. The right lobe of the liver was torn into small pieces, "some of which were loose, and mixed with blood. There was no sign of inflammation in the peritoneum, and the other viscera were healthy." Shell does not comminute bone so much as a rifle ball, but it tears the soft parts much more considerably. To refute the old myth concerning the effects of the wind of a passing ball, calls not even for passing mention in a work of modern times. All the cases of this description of which I heard, were quite explicable on the suppositions laid down by Vacher, in his memoir upon this subject. Under wounds of the head, I have mentioned a case (*Quin*) which would undoubtedly be set down of old as having been so caused. There were many instances of the very near approach and even slight contact of round shot, without any inconvenience arising, further than might be looked for from the unexpected and unwelcome vicinity of such an intruder.

CHAPTER III.

The use of Chloroform in the Crimea—Primary and Secondary Hæmorrhage from Gun-shot Wounds—Tetanus—Gangrene—Erysipelas—Frost-bite.

The advantages derived from the use of anæsthetics are perhaps more evident and more appreciated in the field than in civil practice. The many dreadful injuries which are presented to us in war, and the severe suffering which so often results from them, soon cause us fully to appreciate the benefits bestowed by such "pain-soothers."

The vast majority of the surgeons of the Eastern army were most enthusiastic in the anticipations of what chloroform was to accomplish. It was expected to revolutionize the whole art of surgery. Many operations, hitherto discarded, were now to be performed: and many, which the experience of the Peninsula said were necessary, were henceforth to be done away with.

In the British army chloroform was almost universally employed; but although the French also used it very extensively, as we learn from Baudens, still I do not think, from what I saw of its employment in their hospitals, that they had our confidence in it. Baudens tells us* that "they had no fatal accident to deplore from its use, although during the Eastern campaign chloroform was employed thirty thousand times, or more. In the Crimea alone," "he continues, "it was administered to more than twenty thousand wounded, according to the calculations of M. Scrive."

In one division of our army it was not so commonly used as in the others, from an aversion to it entertained by the principal medical officer of the division—a gentleman of very extensive experience. The only case in which, with any show of fairness, fatal consequences could be said to have followed its use, occurred in the division referred to. The patient, a man of thirty-two years of age, belonged to the 62nd regiment, and was about to have a finger removed. The chloroform was administered on a handkerchief, as he sat in a chair. Death was sudden; and artificial respiration, which was the means of resuscitation employed, failed to restore him. No pathological condition sufficient to account for death was found post-mortem. Some five or six other cases were brought forward by the small body of sur-

---

* Revue des Deux Mondes, Apr. 1857.

geons who were suspicious of the action of chloroform, as having ended fatally from its effects; but in none of these could, I think, the least pretext be found for the imputation, further than that the anæsthetic had been administered at some period previous to death. A man who had been dreadfully mutilated, and who had lost much blood, died shortly after having his thigh removed high up. Chloroform had been used, and to it was ascribed the fatal issue. Death, twenty or thirty hours after a capital operation, rendered necessary by the most dreadful injuries, must be attributed to the chloroform, and so on, and no note taken of the effects of severe injury, *plus* a capital operation, in shattering the already enfeebled powers! Death occurring under such circumstances, when no choloform was employed, would not be thought to demand any special explanation, nor does the fact that the injury was occasioned by a round shot introduce any new element into the calculation.

The objections made to the use of chloroform were restricted to two classes of cases—trivial accidents, in which it was thought unnecessary to run the risk of giving it, and amputations of the thigh, in which a fatal accession of shock was feared. However this may be, it certainly shows the little practical force of these objections, that, while with every indulgence in the interpretation of the law "post hoc," &c., only some half-dozen cases could be obtained throughout the whole army to illustrate the pernicious effects of this agent, and that, too, when thousands upon thousands had been submitted to its action, and hundreds of surgeons of equal experience to the objectors were ready to record their unqualified opinion in its favour, as well as their gratitude for its benefits. For my own part, I never had reason, for one moment, to doubt the unfailing good and universal applicability of chloroform in gun-shot injuries, *if properly administered*. I most conscientiously believe that its use in our army directly saved very many lives—that many operations necessary for this end were performed by its assistance, which could not otherwise have been attempted—that these operations were more successfully, because more carefully, executed—that life was often saved even by the avoidance of pain—the *morale* of the wounded better sustained, and the courage and comfort of the surgeon increased. I think I have seen enough of its effects to conclude, that, if its action is not carried beyond the stage necessary for operation, it does not increase the depression which results from injury, but that, on the contrary, it in many instances supports the strength under operation. Its usefulness is seen in nothing more than when, by its employment, we perform operations close upon the receipt of injury, and thereby, if not entirely, at least in a great degree, are able to ward off that "embranlment" of the nervous system which is otherwise sure to follow, and whose nature we know only by its dire effects.

To men who had lost much blood, it had, of course, to be administered with great care, from the rapidity of its absorption in such persons; but if we do not act on broader principles in its exhibition than reckoning the number of drops which have been employed, or the part of the nervous ystem which we may presume to be at the time engaged, then we must expect disastrous results. It is difficult to see how its use could favour secondary hæmorrhage after operation, as some said it did; but it is, on the contrary, easy to understand how the opposite result might follow. That purulent absorption should prevail among men so broken in health as our men were, need not be explained by the employment of chloroform; and that ice would prove more useful in the slighter operative cases in field practice, few will be disposed to admit, either on the ground of time, efficiency, or opportunity. To Deputy Inspector-General Taylor we owe the practical observation, that chloroform appears to act more efficiently when administered in the open air.*

In the prolonged searches which are sometimes necessary for the extraction of foreign bodies, chloroform is useful, not only in preventing pain, but also in restraining muscular contractions, by which obstacles are thrown in the way of our extraction, which did not oppose themselves to the introduction of the body. Then much is gained in field practice by the mere avoidance of the patient's screams when undergoing operation, as it frequently happens that but a thin partition, a blanket or a few planks, intervene between him who is being operated upon, and those who wait to undergo a like trial. Thus when, as after a general engagement, a vast number of men come in quick succession to be subjected to operation, it is a point of great importance to save them from the depression and dread which the screams and groans of their comrades necessarily produce in them.

It is therefore my clear conviction, that the experience of the late war, as regards chloroform, is unequivocally favourable; that it has shown that chloroform, both directly and indirectly, saves life; that it abates a vast amount of suffering; that its use is as plainly indicated in gun-shot as in other wounds; and that, if administered with equal care, it matters not whether the operation about to be performed be necessitated by a gun-shot wound, or by any of the accidents which occur in civil life.

Hæmorrhage was in the olden time the great bugbear of the military surgeon, and that against which his field arrangements were chiefly directed. It is not now, however, so much feared, from its being well known not to be of so frequent occurrence on the field, and the means of arresting it being better understood.

[*When the vital energies are much depressed by hæmorrhage or otherwise experience has established the propriety of administering an active stimulant before having recourse to chloroform.]

T.

Blandin, in his communication to the Academy, says that his observation of gun-shot wounds led him to believe that primary bleeding always takes place if a vessel of any size is injured, but that it is soon spontaneously arrested by an action similar to that which takes place when a limb is torn off. Sanson repeats the remark as to the constant presence of hæmorrhage to some extent at the moment of injury. Guthrie did good service to surgery by showing how small a force can obstruct a vessel of the first order. He thereby gave courage and confidence to both surgeon and patient.

It has been the experience of most wars, certainly of the late one, that tourniquets are of little use on the battle-field; for though it is unquestionable that a large number of the dead sink from hæmorrhage, still, it would be impossible, amidst the turmoil and danger of the fight, to rescue them in time, the nature of the wounds in most of these causing death very rapidly.* A great artery is shot through, and in a moment the heart has emptied itself by the wound. It would be an experiment of some danger, but of much interest, as bearing on this question, to examine the bodies of the slain immediately after a battle, and carefully record the apparent cause of death in each case. †

I before remarked how curiously arteries escape injury from a ball passing through a limb. In the course of the femoral vessels this is very commonly seen. Through the axilla, through the neck, out and in behind the angles of the jaw, between the bones of the fore-arm, and even of the leg, balls of various sizes and shapes pass without injury to the vessels. Thus, the neck has suffered severe injury many times, and yet very few deaths appear in the returns from these wounds. I have never myself seen any case in which a bullet has passed harmlessly between a large artery and its vein, but many such cases are on record.

The following may be mentioned as instances of narrow escapes. A solder was wounded at Inkerman, by a ball which entered through the right cheek, and escaped behind the angle of the opposite jaw, tearing the parts in such a manner that the

---

* Although this is true as a general rule, yet both Larrey and Colles relate instances in which, by prompt assistance, death was prevented in wounds opening the carotid artery.

† The only mention I have been able to meet with in the records of the medical department of the causes of death on the field, is in a report from the surgeon of the 41st, when serving in Cabul. He mentions, that of four men who were killed outright, three were wounded through the chest, and one through the head. After the contests in Paris in 1830, Meniere tells us that it was a common observation at the Morgue, to which the dead were carried, that the greatest number had been shot through the chest.

great vessels were plainly seen, bare and pulsating, in the wound. Three weeks after admission into hospital he was discharged, never having had a bad symptom. A soldier of the Buffs was wounded in June, 1855, by a rifle ball, which struck him in the nape of the neck. It passed forwards round the right side of the neck, going deeply through the tissues; turning up under the angle of the inferior maxilla, it fractured the superior maxillary and malar bones, destroyed the eye, and escaped, killing a man who was sitting beside him. This patient made a rapid recovery. A French soldier at the Alma was struck obliquely by a rifle ball, near to, but outside, the right nipple the ball passed seemingly quite through the vessels and nerves in the axilla, and escaped behind. His cure was rapid and uninterrupted. Another Frenchman was struck in the trenches by a ball, a little below the middle of the right clavicle. The ball escaped behind, breaking off the upper third of the posterior border of the scapula, and yet he recovered perfectly, without any bleeding taking place. Endless numbers of similar cases are presented to us in military hospitals.

A considerable artery may be fairly cut across, and give no further trouble, beyond the first gush of blood which takes place at the moment of injury. In such cases, the vessel contracts and closes itself. If only half divided, as it is apt to be by shell, or by the quick passage of a ball, then the hæmorrhage will be, in all probability, fatal. The best example, perhaps, on record of the former result, is that mentioned by Larrey. A soldier, struck on the lower third of the thigh by a ball, suffered one severe hæmorrhage, which was never repeated. The limb became cold, the popliteal ceased to beat, and the ends of the divided femoral could be felt retracted when the finger was placed in the wound. This man recovered perfectly. The younger Larrey records a very curious case from the wounded at the siege of Antwerp. A shell passed between a man's thighs, and, destroying the soft parts, divided both femorals; yet there was no hæmorrhage, although the pulsation continued in the upper ends of the vessels to within a few lines of their extremities.

The speed of the ball at the moment when it comes in contact with an artery has a good deal to do with the injury it inflicts. If it be in full flight it may so cut open the vessel as to allow of instantaneous hæmorrhage; whereas, if its speed be much diminished, the contusion it occasions opposes immediate, but favours secondary bleeding.

Primary hæmorrhage may take place either instantaneously on the receipt of a wound, or after a little time, when the faintness resulting from the accident has gone off. I have already referred to some instances in which the former is liable to occur. In wounds of the face, too, this instantaneous bleeding is very usual.

Some cases occurred in the Crimea of the well-known fact,

that limbs may be carried away, and their arteries hang loosely from the shattered stump without bleeding. Two came under my own notice, in which legs were carried away by round shot, and no hæmorrhage took place, though both men died subsequently from other causes. This spontaneous cessation of hæmorrhage is perhaps most commonly seen in the upper arm.

The returns fail to inform us of the number of cases either absolutely or proportionately to the whole number of wounds, in which secondary hæmorrhage took place during the war. Although I have no figures to which I can refer as corroborating the statement, yet I am inclined to think that the proportion of cases in which serious bleeding did take place, is higher than that set down by Mr. Guthrie. The distinction drawn by Dr. John Thomson between secondary hæmorrhage proceeding from sloughing, ulceration, and excited arterial action as it occurs at different stages of treatment, is a good one. That which takes place after twenty-four hours and up to the tenth day being usually due to sloughing, resulting directly from the injury, should always have the term "intermediary" applied to it; and the bleeding which proceeds from morbid action, such as ulceration attacking the part, and which takes place at a later period, would be more appropriately called "par excellence" secondary. Hæmorrhage should thus be distinguished into three periods: "primary," occurring within twenty-four hours; "intermediary," between that and the tenth day; and "secondary," that which takes place at a later date. More precision would be given to our language on this important subject, by such a distinction being always recognized.

The period at which consecutive bleeding is most apt to take place has been variously estimated. Guthrie sets it down as occurring from the eighth to the twentieth day, Dupuytren from the tenth to the twentieth, Henman from the fifth to eleventh, and Roux from the sixth to the twentieth. In the cases I have myself observed, it has taken place between the fifth and twenty-fifth days, and by a curious coincidence, it has appeared in the majority on the fifteenth after the receipt of the wound. In one case, a wound without fracture of the thigh, it was said to have taken place as late as the seventh week, and that when no gangrene or apparent ulceration was present.

Consecutive hæmorrhage may occur from very insignificant vessels, and be arrested by simple means; but when it takes place from a large arterial trunk, it is an accident of the most serious importance.* With us, in particular, such effusions were

---

* The following is a very interesting case of secondary hæmorrhage caused by the limited ulceration of a large artery, which is related by Dr. Scott of the 32nd, in a report existing in the archives of the medical department:—' Private John Hodgson, age 31, was

causes of extreme anxiety, as the deteriorated state of the health of our patients made such an accident peculiarly disastrous. Their strength could not withstand such a drain, and the scurvy made their blood so thin and effusible that they were liable to great loss of blood, not by vigorous hæmorrhages, but by slow, though not less destructive, discharges. From this it can be understood that in the Crimea many of the time-honoured remedies for hæmorrhage, such as venesection, starving, &c., were entirely discarded, and replaced most generally by their opposites. Tonics, as quinine and iron, were the remedies most wanted; and as to styptics given internally, they always appeared to me to be mere farces, except in so far as they acted as general tonics.

The more useful prophylactics to such consecutive hæmorrhages, such as quiet of mind, and perfect rest of the wounded part, are not always attainable in field practice, especially when the necessity of removing patients occurs so frequently. It is of course impossible altogether to avoid such movements during war, but it is most unfortunate that they fall so often to be executed at the very period when they become most dangerous. No man, at all severely wounded by gun-shot, can be considered safe from hæmorrhage till his wound is closed, but yet, after twenty-five days, the danger may be said to be in a great measure overcome. In reference to this point a siege has an advantage over an open campaign, from the greater fixedness of the hospitals, and the less frequent moving.

Hæmorrhage occurring early was universally treated by the rule laid down by Bell and Guthrie, of *tying both ends of the bleeding vessel.* When, however, the bleeding appears at a late date, when the limb is much swollen, its tissues infiltrated, matted together, and disorganized, it is by no means an easy thing to follow this practice. The difficulty is perhaps greatest in wounds of the calf of the leg, where the muscles are much developed, when the posterior tibial has repeatedly bled, the wound large and irregular, the contusion severe, and the blood welling out from among the disorganized tissues in no collected stream. The rules and precepts laid down in books about the

---

struck by a ball at Mooltan, about a line anterior to the left carotid artery, below where it divides into the external and internal, and passing through the œsophagus, escaped at a point corresponding to its entrance. No unfavourable symptom appeared for nine days, when a fit of coughing came on, and blood issued from both the mouth and the wounds, and the patient instantly expired. The *right* carotid had been grazed at its bifurcation, and a piece of it about the size of a small pea, and including all its coasts, had sphacelated, and, giving way, caused death before assistance could be got."

appearance of the vessel and the orifice, about the mode of passing a probe towards it from the surface, and the best way of cutting so as to fall upon the vessel, are all worse than useless; as they lead us to expect guides where there are none, but those which watchful eyes and careful incisions afford.

From the results of several cases which fell under my observation in the East, I have reason to believe in the soundness of the views lately put forth by Nelaton, in opposition to the long-credited opinion of Dupuytren, as to the unsound state of the artery in suppurating wounds. I feel pretty sure that the vessel will, in most cases, bear a ligature for a sufficient time to fulfil the end we have in view in its application. It will be necessary to attach it with caution, to employ no more force than is absolutely necessary, and we may expect it to separate, as Nelaton shows, before the usual time, yet it will continue attached sufficiently long to close the vessel, if we do not keep pulling at it so as to tear it away prematurely. It requires but a small force to oppose the blood-impulse, and that the vessel will commonly stand, if carefully handled.

The French, although generally applying the ligature at the seat of injury in primary hæmorrhage, perform Anel's operation when the bleeding appears late. The teaching of Dupuytren and Roux has done much to prevent "the English practice" being so fully followed as it is with us.*

---

* M. Roux, in the second volume of his recently published posthumous works, thus sums up his experience on secondary hæmorrhage from gun-shot wounds. It proceeds, he says, from (1) separation of the eschar; (2) from injury by fractured bones; (3) from the capillaries caused by general feebleness in the patient; (4) hæmorrhage from the erosion or tearing of a vessel, appears later than that arising from the separation of the eschar, the one appearing about the eighth or tenth day, and the other from the fifteenth to the twentieth; (5) hæmorrhage arising from the tearing of a vessel, and especially that which accompanies compound fracture, is more common in wounds of the thigh than any other; (6) whatever be its cause, the manifestation of the bleeding is very uncertain, being sometimes preceded by symptoms which announce its approach, and sometimes giving no indications of its coming—sometimes it appears in large quantities, and very suddenly, while at other times it appears in small quantities, and will often recur if no interference be had recourse to; (7) sometimes the bleeding takes place within the limb, where it forms a sort of false aneurism, but at other times it flows freely outwards; (8) when the bleeding vessel can be got at, we should ligature it, or the trunk from which it proceeds; (9) here, as in the case of all traumatic hæmorrhages, primary or secondary, it is best to tie the vessel in the wound, above and below the place of injury; in general, however, it will be necessary to ligature the vessel at a distance on the distal side of the

Anel's operation is undoubtedly the best in one class of cases dwelt upon by Dupuytren, in which hæmorrhage arises from the tearing of an artery in a simple fracture. The ligature of the main vessel commonly succeeds in these cases, while to find the bleeding vessel is most difficult, and to expose the seat of fracture to the air is a risk greater than should be encountered.

There are, unquestionably, some situations where it is impossible to get at the wounded vessel, especially when the bleeding has taken place at a late date. The deep branches of the carotid afford, perhaps, the most patent example. In a case of this sort, in which the deep temporal and internal maxillary were wounded, in a Russian admitted into the general hospital after the assault in September, Mr. Maunder tied the carotid to arrest the bleeding, which had recurred several times, notwithstanding pressure. The ligature of the main vessel commanded the hæmorrhage, although the patient subsequently died of exhaustion.

Secondary hæmorrhage may appear at a very late date from ulceration, set up by the pressure of a fragment of bone pressing upon the vessel. The ulcerative process in these cases is sometimes very slow.

The following case is interesting, and conveys much instruction as to the value of the different places in which to apply the ligature. A Russian boy who had sustained a compound fracture of the leg at Inkerman, from gun-shot, was received into the French

---

wound, after the methods of Anel or Hunter, because the difficulties of finding it are great, and its state in the wound will not allow of a ligature being applied to it there.

Sanson, again (Des Hæmorrhagies Traumatiques), concludes thus: "A ligature applied to the two ends of a divided artery, is the surest method of arresting the bleeding, and to prevent a return. But we do not think, after the example of the English surgeons, that it should be put in force in all cases, and whatever be the situation of the artery, from the risk of causing great destruction, violent inflammation, and long suppurations. We often meet with wounds attended with hæmorrhage, or false primitive aneurisms, in which it is difficult or impossible to determine which is the divided vessel. In other cases we recognise the source of the bleeding, but it is situated too deeply for us, without causing grave injury, to find and tie the artery above and below the wound. We are thus compelled to ligature this artery, or at least the trunk from which it proceeds, between the heart and the wound, but at a considerable distance from the latter. It is true that traumatic hæmorrhages are much less favourable than aneurisms, properly so called, to the success of Anel's method. But it is a necessity in a way, and besides we can, if the method of Anel fails, have recourse at a later period to the ligature of the two ends in those cases in which it is possible."

hospital at Pera a few days afterwards. On the fifteenth day from the date of injury, profuse hæmorrhage took place from both openings. Pressure failed to arrest it. The popliteal was tied the same day, according to the method recommended by M. Robert, viz.: on the inner side of the limb, between the vastus and hamstring muscles. The foot remained very cold for four days, and then violent reaction set in; and on the eighth day from the ligature of the main vessel hæmorrhage recurred, both from the original wound and the incision of ligature. Pressure was again tried in vain. The superficial femoral was next ligatured, on the tenth day from the deligation of the popliteal. Four days afterwards the bleeding returned from the wound, and pressure then seemed to check it. The ligature separated from the femoral on the twelfth day after its being applied, and the third day after, *i. e.*, the twenty-fifth day from the first occurrence of the hæmorrhage, bleeding having again set in from the wound, the limb was amputated high in the thigh, and the unfortunate patient ultimately recovered. Would Mr. Guthrie not have saved this man's limb, and the surgeons much trouble?

In gun-shot wounds of regions where the vascular communications are at all free, the ligature of the main trunk for consecutive bleeding cannot often be of any use, as it is seldom possible to be sure that the hæmorrhage proceeds from the main vessel, nor yet can we by such an operation cut off the collateral circulation. If the source of the hæmorrhage could be certainly ascertained, and if pressure could be applied to the lower portion of the divided vessel at the same time, then we might reasonably hope to arrest the bleeding by tying the main artery; but the mere placing of a ligature on the proximal side can give no security against the continuance of the bleeding. If the sloughing preceding, and accompanying the bleeding, be extensive, and situated in a muscular and vascular part like the calf of the leg, and if the hæmorrhage has continued notwithstanding the employment of means applied locally, I should never hesitate between amputation and ligature of the main trunk, but have instant recourse to the former, as being the only reliable and satisfactory proceeding.* The following may be taken as a good

---

* I may note the following figures, in passing, as a small contribution to the statistics of this question. The French, in one hospital at Constantinople, ligatured the femoral at a distance from the wound for secondary hæmorrhage seven times, and all failed. The subclavian was ligatured under like circumstances once, and it succeeded. I have found the detailed report of only four cases in which the main vessel was tied in India. The ligature was applied twice to the femoral, once to the brachial, and once to the radial. It succeeded in arresting the hæmorrhage in three cases; one femoral failed. Dupuytren ligatured the femoral several times

example of a class of cases frequently occurring. M'Gartland, a soldier of the 38th regiment, an unhealthy man, who was still suffering from the effects of scurvy and fever, was shot from the outside and behind, forwards and inwards through the left leg, on the 18th of June. The fibula was broken, and the edge of the tibia was injured. He walked to the rear without assistance. On admission into the hospital, the limb was greatly swollen. This swelling, by appropriate means, very much diminished in a few days. On the fifth day arterial bleeding, to a limited extent, took place from both openings. Recalling a case put on record by Mr. Butcher of Dublin, of a wound of the post tibial, I determined on trying the effects of well-applied pressure along the course of the popliteal, and in the wound, combined with cold and elevation. The limb was also fixed on a splint. The object of the pressure on the main vessel was to diminish, not arrest, the flow of blood through it. On the eighth day there was again some oozing. Pus had accumulated among the muscles of the calf, and required incision for its evacuation. On the ninth day a pulsating tumour was observed on the external aspect of the leg, "the consecutive false aneurism" of Foubert, and next day the bleeding returned from both wounds. I wished then to cut down and tie the vessel in the wound, but a consultation decided on waiting a little longer, in the hope that the bleeding might not return. On the night of the eleventh day most profuse hæmorrhage recurred. The attendant, though strictly enjoined to tighten the tourniquet, failed to do so, but the necessary steps to arrest the bleeding were taken by the officer on duty. Next morning, when I first heard of the occurrence, I found the patient blanched, cold, and nearly pulseless. A consultation decided that the state of the parts made the securing of the vessel in the wound very problematical, and that, as the limb would not recover if the main artery was taken up, amputation must be performed so soon as the patient had rallied. When reaction had fairly taken place, I amputated the limb. The removed parts were much engorged, sloughed, and disorganized. The anterior tibial was found to have been opened for about an inch shortly after its origin, and on it was formed the aneurism, which had a communication with both orifices of the wound.

In all such cases the second bleeding should determine active interference. I say the second bleeding, as it very often happens, that when hæmorrhage has taken place once, even to a considerable extent, and evidently from a vessel of large calibre, it never recurs. Many most striking instances of this have come

---

for bleeding from the calf, but with what result it is impossible always to make out. S. Cooper, while he once successfully took up the popliteal for secondary hæmorrhage from a wound of the posterior tibial, strongly reprehends the practice as a general rule.

under my notice. But though more than even this is true, and
that frequently blood thrown out repeatedly is spontaneously arrested, still the great preponderance of cases, in which it recurs
in dangerous repetitions and quantities, as in the above instance,
should cause us, I believe, to interfere on its second appearance,
more particularly if the bleeding be in any quantity. Not to
interfere unless the vessel is bleeding, must not always be understood too literally, or we will often be prevented from performing
the operation till our patient is beyond our help. The hæmorrhage recurs over and over again, and the surgeon, though as
near as is practicable, arrives only in time to see the bed drenched,
and the patient and attendant intensely alarmed. There is at
the moment no bleeding, and he vainly hopes there will be no
return; and so on goes the game between ebbing life and menac-
ing death, the loss not great at each time, but mighty in its sum,
till all assistance is useless. Many a valuable life has thus been
lost which might have been saved by a more decided course of
action.

Few cases are more embarrassing than those to the surgeon,
or require more determination, and well-considered resolution to
conduct to a successful issue. One is averse to act when the immediate necessity has passed; and unless we be guided in our
course by a knowledge of general results, more than by the immediate case in hand, we will lose many a patient. These
cases form an exception to the rational surgery of the day,
which prescribes inaction, unless there be immediate call for
interference. There can be little doubt that hæmorrhage may
often be definitely arrested by pressure applied with care along
an extensive part of the wounded artery, as well as to the
apertures; but such treatment is not adapted for gun-shot
wounds, from the depth and narrowness of their tracks, unless
we so enlarge them as to admit the compress deep into the
wound. This was shown in the case recorded above, as well as
in many others. The discharge is pent up by the plug, and burrows largely among the tissues.

There were many cases of hæmorrhage from the hand succeeding gun-shot wounds, which came under my notice during
the war. Many of the injuries resulted from the accidental
explosion of the patient's own gun, and, I suspect, in not a few
cases from inattention. Hæmorrhage, in such instances, was at
times very troublesome, especially when the bones of the hand
were much fractured, as it was then difficult, if not impossible,
to secure the vessel in the wound. The secondary bleeding
usually appeared early in these cases, and, so far as my observation went, ligature of the brachial seems better practice when
local means fail, than putting a thread on the vessels of the fore-
arm, as I saw done several times in the East. In recent cases
we can often ligature the bleeding vessel, but in the sloughing
stage, with a deep wound, and the bones much injured, it is im-

possible to secure it. To ligature the radial and ulnar separately, or conjointly, exposes the patient to operative dangers which bring no adequate return, as the probability of success is very small. In the following case, the ligature of the vessels of the forearm succeeded; but in four other cases, in which I knew it tried for wounds of the palm, it failed utterly. The position of the wound in this case made it more likely that the proceeding followed should succeed:—A soldier, resting his right hand on his musket, was struck by a ball on the web between the thumb and forefinger. The wound seemed trivial, but the whole hand swelled exceedingly. On the fourteenth day arterial hæmorrhage occurred, and pressure was applied. The bleeding repeatedly recurred, and still pressure was preservered in. Finally, the radial, and then the ulnar, were ligatured before the hæmorrhage was commanded. An early search in the wound, would probably have succeeded in securing the vessel.

I have seen the method of pressure on the brachial by flexing the arm, and by bandaging; both fail in some of these cases.

Hæmorrhage from the face of stumps is unquestionably one of the most disagreeable complications which can arise during their treatment. The scorbutic state of the blood of most of our men, made their stumps highly irritable, and liable to sanguinolent oozing. Their strength was thus much wasted, and other complications of hardly less serious importance were superinduced. All noticed the prevalence of these bleedings when the hot sirocco blew, or previous to those violent thunder-storms which did so much to clear the air. The patients often complained at these times of feeling "as if their stumps would burst," and the bleeding seemed to give them much relief. The blood which flowed was commonly more venous than arterial, thin, watery, and of a brick-dust colour. When cold air, or water, combined with elevation, failed to check it, pressure and the perchloride of iron generally succeeded. Its appearance was always an indication for more fresh air, tonics, and better food.*

Guthrie counsels us, in the event of hæmorrhage from a thigh stump which cannot be commanded by the application of a ligature to the bleeding point, to tie the main vessel, in the first instance, at a point the nearest to the end of the stump, at which pressure commands it, provided it be beyond the sphere of the

---

* Briot (Hist. de la Chir. Milit.) remarks, that strong vigorous subjects are not those in whom he has seen hæmorrhage, either primary or secondary, most commonly follow gun-shot wounds; but, on the contrary, it was more common in patients of an opposite character. This he ascribes to the want of tone in these men preventing the contraction, or closure of the vessels. The same thing, he says, exists in the power we have of arresting bleeding in primary and secondary operations—those necessitated by accident and disease.

inflammation; and if this fail, then to reamputate the limb. He adds, that if pressure above the going off of the profunda is necessary to command the bleeding, then we should amputate, in place of tying the vessel in the groin. The dictates of so great a master are not lightly to be controverted, but, so far as my comparatively very limited observation goes, I would be disposed to tie the iliac, rather than either ligature the femoral high up, or reamputate the limb; this is, of course, always providing that the vessel could not be secured on the face of the stump. Bleeding from a thigh stump is so apt to proceed from one of the deep vessels, and to be temporarily arrested by a tourniquet applied to the femoral, but whose strap encircles the limb, that no ligature of the femoral much above the extremity of the stump could give any security against a return. The fear of gangrene in depressed subjects when the iliac is tied, is the chief objection to the practice I allude to.

The cases in which attempts were made in the East to arrest hæmorrhage from stumps, by applying a ligature to the main vessel above the extremity of the stump, were, I believe, singularly unfortunate. Well-applied pressure along the course of the principal vessel, adapted to diminish the circulation through it, has sufficed, in some few most threatening cases, finally to arrest bleedings which had recurred frequently; but in these cases the implication of the main vessel was clearly made out. Take the following case as an example. The state of the vessel, as discovered after death, also lends an interest to the narrative. Hæmorrhage took place to a slight extent from a thigh stump, on the ninth day after operation, and was repeated on the following morning. A tourniquet was applied over the course of the femoral, so as to moderate the flow of blood through it. On the fourteenth day the bleeding returned, and the tourniquet was tightened for four hours, so as almost to arrest the current of blood in the great vessel, and afterwards, though loosened, was still left so tight as to restrain the free flow of the blood through the main artery. On the sixteenth day the bleeding returned, and the same treatment was followed, the position of the compressing force being carefully shifted from time to time. From this period the hæmorrhage never reappeared. The patient subsequently died of pyœmia, when it was found that an abscess had formed around the great vessels, extending from the end of the stump upwards for some inches; that the artery was fairly opened by ulceration, to the extent of an inch from its termination, but beyond that distance a dense clot occupied its calibre for an inch and a quarter. The vein contained much pus. Purulent matter was freely deposited in the lungs. Here the ulceration of the end of the artery allowed the bleeding to take place, while the subsequent formation of a clot above arrested it.

The exact number of cases in which *tetanus* has followed wounds during the war, or the nature of the injuries giving rise to it, I have failed to learn from the army returns. It was not, however, by any means common. I know of six cases only which occurred in camp, and seven which took place at Scutari. The usual proportion, to wounds is, according to Alcock, one in seventy-nine. We have certainly not had that ratio. In no case, the particulars of which I could learn, did it occur after the twenty-second day—the limit as defined by Sir James M'Gregor. In the cases of which I have known the details, there was no confirmation of Baron Larry's theory as to the set of muscles affected according to the position of the wound. The cases occurring in our army have been, so far as I know, with one exception, universally fatal. Of the six instances which appeared in front, one followed a compound fracture of the thigh, one a face wound with destruction of the eye, one an amputation of the shoulder, one a flesh wound of the leg, and the other two cases following wounds, without fracture of the thigh, unfortunately happened under my own charge in neighbouring wards in the general hospital, and within a very short time of one another. Of those which appeared at Scutari, one followed an amputation of the hand, two succeeded compound fractures of the thigh, one was a frost-bite of the toes, one was a compound fracture of the leg, and of the other two cases I could not learn the primary lesion. I give the particulars of my own cases, from their presenting some points of interest, of which not the least was their extreme similarity to one another.

Hughes, a private of the 44th regiment, was admitted into the general hospital on the 18th of June. In the assault on the Redan, a ball had entered an inch below the anterior superior spinious process of the right ilium, and passing downwards and outwards, lodged deeply among the muscles of the thigh. After a most careful and prolonged examination, its position could not be ascertained, and it was left, in the hope that in a day or two it might become defined, or that it might perhaps remain altogether without doing harm. There was no fracture, and no pain. The case was treated as a flesh wound. On the thirteenth day, the patient for the first time complained of pain behind the great trochanter of the right side, and the presence there of deep fluctuation caused me to make an incision. A considerable accumulation of pus was found, and in the sack of this collection I discovered the ball much flattened. I freely enlarged the wound, so that all retention of matter was prevented. Next day some cloth was discharged from this opening. On the seventeenth day his manner was changed. He was irritable, and complained of his wound. *Pus continued to flow freely, and his general health was unimpaired.* He said that he had caught cold, and "that it had taken him in the jaws," which were a little stiff. His bowels being costive, I ordered him a purge, and an embrocation for his jaws. I had not at this time any suspicion of the impending evil. Next day I found that his bowels had been fully moved, and that the most offensive dark-coloured stools had resulted. The trismus was now very marked. The masseters were hard and contracted, like clamps of iron. I examined the wound, and further enlarged it. A large, emollient cataplasm was applied, and a drop of croton oil given internally. His bowels were freed of much more of the same fetid dejection which he had voided formerly. In the afternoon, when rising to go to stool (which he insisted on doing), he had a violent spasm over the right side of the body, not accompanied with any pain. From this time the spasmodic contractions set in, recurring at certain intervals, leaving him at times for half a day, but always returning till his death. These spasms were nearly confined to the wounded side, and affected the muscles of the thigh most. I began the use of the acetate of morphia in gr. ij. doses, and afterwards diminished it to one grain administered every hour till he slept. This he did in snatches during the succeeding days, waking up startled if any one walked near his bed. Whenever he slept for an hour or two, his symptoms were markedly alleviated. When he slept, the opium was intermitted for some hours, and then resumed. Only on the first day did he exhibit the slightest symptoms of narcotism, and so much relief did he experience from the use of the drug, that he earnestly asked for it whenever he was a few hours without sleep. He always denied suffering any pain, though from the way in which the corners of his mouth was drawn upwards and backwards, so as to expose his teeth, and the manner in which his brow was knit, he looked as if he was in extreme agony. There was one small spot, presenting no peculiarity to the eye or touch, on the inner side of the knee, and another on the ankle of the wounded limb, which he always said gave him much pain. The least pressure on these spots always caused the most violent spasm. He frequently expressed his astonishment at the limb starting in the way it did, and tried in vain to prevent it. His mind remained unaffected till near his death, when he became dull and heavy like a drunken man. The muscles of the neck, the long muscles of the back, as well as the *serratus magnus* of the right side, and the muscles of the thigh and leg of the same side, became hard as a board, particularly during the transit of a spasm. So hard and contracted were they, that when we had occasion to move him in bed, he could be raised like a log of wood, at least so far as his right side was concerned. His abdomen was much distended, and its muscles hard. The clysters which were administered during the course of his treatment, always give him much relief from the feeling of "bursting," of which he so often complained. He lay diagonally in the bed, his wounded limb stretched straight out, and the other drawn up. Latterly he suffered from a severe pain, which continued to shoot from the ensiform cartilage to his spine, and also from intermitting pains in his right side. For a couple of days there was a diminution of the discharge from the wound, but ultimately it became quite re-established. His skin was always bathed in perspiration, the excretion having a most pungent and offensive smell. For some days before death a miliary eruption showed itself over the upper part of his body. His pulse was slightly elevated during the course of the malady, but it never reached any very high standard. His respirations varied from twenty-six to twenty-eight per minute. A very viscid spittle, which he was always trying to hawk up, gave him much annoyance. He had retention of urine, and latterly suppression. There was some blood mixed with his urine for a couple of days.

I have already alluded to the treatment which was followed. Purgatives, opium given freely, at first combined with camphor, and latterly alone. He frequently took as much as fifteen grains of opium before he slept, and altogether he must have consumed a great quantity of that drug. He asked for fomentations to be applied to his limb, which to the hand felt colder than its fellow. Their application, he said, gave him relief. His ability to swallow semi-fluid food enabled me to give him the most nutritious diet I could devise, along with wine. With intermissions and exacerbations his fatal malady progressed. A spasmodic cough was added to his other ailments. On the afternoon of the tenth day of attack, his symptoms greatly abated for some hours, and while he was conversing with a comrade, he was seized with what the orderly termed "a fit," vomited some dark matter, was severely convulsed, so that the body was drawn backwards and to the right side, and before I reached his bedside, he was dead. By a mistake of the hospital sergeant, no post-mortem examination was got. I looked hurriedly at the wound shortly after death. The fascia lata was much lacerated, and the parts beneath were sloughy.

Barker, a private in the 38th regiment, aged 20, was admitted into the general hospital in camp on the same day as the last patient, June 18th. A ball had penetrated his left thigh at its inner and lower aspect, and lodged. It could not be found, though every means were used. Four days afterwards it was felt near the wound, and removed. By the 28th the wound was looking sloughy, and the discharge was thin and unhealthy. He complained much more than was usual about his wound, and appeared very anxious. On the 30th I noticed some twitching of the limb as it was being dressed. His bowels were free, but he complained of sleeping little at night. The wound was freely enlarged, and covered with a poultice. He was purged with croton oil and clysters. He grew gradually worse. During the two succeeding days, the spasms were very decidedly pronounced over the left side. He described them himself as proceeding in "flashes" from his wound to the spine, and back again. Touching the limb, and particularly the sole of the foot, immediately aroused the most violent spasmodic contractions. His pulse rose to 92, and his respirations to 29 per minute. He did not complain of pain, but was greatly distressed by a thick spit which clung to his teeth, and which he was always making violent attempts to expel. The left side of the body was almost alone affected, and the spasms, as in the last case, drew him diagonally backwards, and to the wounded side. He had no trismus for the first day, but afterwards it became marked. He always said that he was sure, if he could only sleep, he would be all right. I brought him under the influence of chloroform, and while its effects continued, the spasms were relieved, and certainly the pulse and respirations were reduced in frequency; but so soon as he awoke, all his worst symptoms returned in undiminished vigour. Having seen the utter futility of chloroform to relieve the spasms permanently, or to arrest the disease, in two former cases at home, where the anæsthetic had been fairly tried; and having many wounded to attend to, and no assistant to whom I could intrust the exhibition of the anæsthetic, I determined to abandon it and trust to opium. This, with enemas, nourishing food, and local emollient applications, comprehended all the treatment. The symptoms were not abated, except for short intervals, and then only in proportion as sleep was procured. His skin was always covered with an odorous perspiration. The abdomen got distended and hard. The muscles of the back were markedly hard and contracted, particularly on the left side. The left leg was stretched out spasmodically, every muscle defined. The right limb was drawn up, and he lay across the bed. The wound was sloughy, and shreds of fascia escaped with the discharge. The urine became scanty and high-coloured, and required to be drawn off by the catheter. Eventually he suffered much pain in the left groin and calf of the leg, as well as at the ensiform cartilage. When trying to raise himself on his elbow on the fifth day of the attack, and seventeenth after admission, he was violently convulsed, so that he was bent greatly backwards; he put his hand to his throat as if choking, and fell back dead.

The wound was found to be lined with an ashy slough. The bone was not injured. The fascia lata was much torn, and was pierced and ulcerated at a spot on the anterior and external aspect of the limb, some little distance from the wound. The ball had evidently penetrated to this point. No nerve fibres could be detected near the wound. The parts in the neighbourhood were sound. The brain and internal organs were healthy. The lungs were only slightly congested, and viscid mucus was present in the larger tubes. The spinal canal contained a good deal of fluid blood. The cord and its membranes were congested. In the lower cervical and upper dorsal region the substance of the cord was varicose—contracted and expanded into a series of knots. There was no other pathological appearance.

On looking at these two cases in connection, the curious parallelism must strike one. The very distinct manner, too, in which so many of the peculiar symptoms of this deadly disease were developed, particularly in the first case, was interesting. Whether opium, whi ɩ appeared to act so beneficially, had it been pushed further, so as to produce a more decided impression, would have done good, is a question. I believe it would have affected the result but little. The similarity between the wounds in these two cases, the non-discovery of the balls for some days, the symptoms, the season of the year, and the state of the cord in the last case, were all interesting. That the high temperature we had at that period had much to do with the production of the disease, is not certain; yet three of the six cases which occurred in front appeared during a period of extreme heat. In one case of tetanus, succeeding an injury of the foot, which recovered, chloroform was repeatedly administered for prolonged periods, and anodynes applied to the spine. The particulars of this case are, I understand, to be published by the surgeon in charge, Dr. Ward of the 17th regiment.

As to treatment we are yet unfortunately in the dark. Romberg sums up his review of the question thus: "The results of treatment amount to this, that wherever tetanus puts on the acute form, no curative proceeding will avail; whilst in the milder and more tardy form, the most various remedies have been followed by a cure." Larrey trusted most to opium and camphor, with section of the nerve in cases adapted for it. On reading the many accounts which have been given of cases of this disease, opium and chloroform appear decidedly to have the greatest evidence in their favour.

The unpublished records of the Indian campaigns, illustrate to a great extent the remarkable effect which unextracted balls seem to exercise on the development of this fatal affection, more especially when they lay under strong fasciæ, as in my cases. In India, as well as in the continental campaigns, amputation at the shoulder appears to be one of the operations most frequently followed by tetanus.

Sudden vicissitudes of temperature have been always looked upon as most powerful causes of tetanus, especially the change from a hot day to a cold and damp night, which is so common in the tropics. So it was after the battles of Ferozepore and Chillianwallah, when the wounded lay exposed to very cold nights succeeding days of hard work under a burning sun. Larrey notices the same circumstances as having predisposed to the disease in Egypt, and in the German campaign of 1809. After Bautzen the exposure to a very cold night produced over a hundred cases, and after the battle of Dresden, when the wounded were placed in like circumstances, they lost a very large number from tetanus. Baudens gives a very interesting recital from his African experience, which shows the influence of cold and moisture in producing this disease. Forty slightly wounded men were placed, in the month of December, and during the prevalence of a north east wind, in a gallery on the ground floor, which was open to the north. Fifteen different cases of tetanus appeared in a short time; among this number twelve died. The remainder were removed to a more sheltered place, and there were no more attacked. The exposure after the Alma might have been expected to produce many cases; but I do not believe that many resulted therefrom, though the confusion which existed, with regard to reports, at that period, makes it difficult to know what was the real effect of such exposure in reference to this point.

Opposite extremes of temperature appear to cause similar effects in this most curious affection. In both the Indies, heat is looked upon as a most powerful predisposing and exciting cause, and idiopathic tetanus is there not uncommon both among the natives and the European troops, while in the arctic region it is even more frequent and fatal. Sir Gilbert Blane tells us, that out of 810 wounded men who came under his observation in the West Indies in 1782, thirty were seized with tetanus and seventeen died. Dr. Kane's experience in the arctic regions, shows how apt exposure to a low temperature is to cause it. He tells us, that while most of his party were more or less affected, he lost two men from an "anomalous spasmodic affection allied to tetanus," and that all his dogs perished from a like cause. The great cold, exposure, and frost bites, which were sustained in the Crimea during the first winter, were followed by fewer cases of tetanus than we might have expected, though I suspect more cases appeared than we have any record of.

I am ignorant of the total number of cases which have occurred in the French hospitals; but of five cases with the history of which I was familiar, and which appeared about the same period in the hospitals at Constantinople, one followed compound fracture of the thigh; two, wounds of the foot; and one, a penetrating wound of the chest in a Zouave, who, after recovery, was allowed to visit the city, where he remained drunk for three days: he was seized with tetanus on his return, and died in forty-eight hours. The French trust mostly to opium in the treatment, and report favourable results from its use; though I suspect from what I have heard, that not a few cases of simple trismus were inadvertently classed by them under the more formidable disease of tetanus.

I have been put in possession of the particulars of some cases which occurred in India, where amputation was had recourse to in tetanus. They all ended fatally without relief, though if performed early, before the peripheral irritation had time to set up much centric disturbance, this step would certainly appear to promise good results, in so far as the cause being removed, local applications to the spine would have a better chance of succeeding in allaying the excited action.

Hospital gangrene was not common in the East. During the first winter it prevailed a good deal in a mild form at Scutari, but it never became either general or severe. It did not appear to pass from bed to bed, but rose sporadically over the hospitals. It frequently attacked the openings both of entrance and exit, but occasionally seized on one only, showing apparently a predilection for the wound of exit. At times it showed itself only in part of a wound, and spread in one direction alone. It was never severe, and was invariably, as far as I saw, of the variety designated "ulcerous" by Delpech, and "phagedæna gangrenosa" by Boggie. In many cases the best designation for it, as it appeared with us, would have been the old one of "putrid degeneration." The earliest symptom was pain in the part which sometimes preceded the ulcerative process by a couple of days. The edges of the wound did not swell up, but remained thin as they were undermined. The pain generally continued during the process of destruction. It appeared chiefly in the lower extremities, and in wounds whose progress towards cure had been for some time stationary. It seldom burrowed far into the intermuscular tissue, but confined its ravages to the surface and the circumference of the wound. I never saw any marked gastric disturbance attend it. If it attacked the wounds of those already labouring under fever, it appeared to aggravate the fever.

The abominable state in which the barrack hospital at Scutari was during its early occupation, may well have caused an outbreak of hospital gangrene among the broken-down men who lay so thickly around the doors of the offensive latrines; but I cannot say that I noticed any greater tendency to its appearance at these places than in any other portion of the hospital. The corridors presented, I think, the greatest number of cases. Whenever it appeared the patients were isolated, and sent into wards set apart for the treatment of the disease.*

Nitric acid, applied locally, and the exhibition of the tincture of the muriate of iron internally, in half-drachm doses, three times daily, proved to be the most efficacious means of stopping it as it appeared in our hospitals. The local nature of the complaint was universally recognized, and local measures relied on for its relief. The application of the escarotic, not only to the edges of the sore, but also to the healthy tissues, at a little distance round the margin, secured by far the best means of employ-

---

* How far this segregation into separate wards is a good plan, I am disposed to doubt. That the malignancy of the disease is thereby increased, and the danger to the other inmates of the hospital enhanced, has been the opinion generally held on the adoption of such measures. If each patient was taken outside of the hospital, and placed in a tent by himself, it would be the most successful way of treating such cases. In an outbreak of this kind, wooden huts would be found most excellent hospitals, as I know from experience elsewhere.

ing the remedy. A barrier of lymph appeared to be thus thrown up around, which prevented the spread of the peculiar inflammatory or destructive action in the skin and cellular tissue to which it was always confined. The attendant fever was uncertain in its development; sometimes it preceded, sometimes it accompanied, and sometimes it followed the local outbreak. Often there was little, if any constitutional disturbance, and occasionally the fever was of a low typhoid type. The most generous diet was always necessary; for though it may be true, as was the case in the Peninsula, that an antiphlogistic treatment is at times necessary, it can be so only in strong healthy men, who derived the disease from infection. With us the depression of the powers of life was so marked, and appeared to exercise so strong an influence, as predisposing to its outbreak, that, in place of lowering remedies, the most strengthening, including stimulants, and above all fresh air, were absolutely required, and were alone of any use.

Those who had suffered in camp from diarrhœa, and whose strength had thus been much reduced, more especially those whose constitutions were strongly impregnated with scurvy, were most liable to be attacked; and, in all our cases, so far as I saw, the development of the disease resulted from a lowered general health more than from specific causes. It was, in many cases, a veritable "child of the typhus." The peculiar dark hue of the face, spoken of by writers, was not common, though it was occasionally seen; but the disagreeable smell and the rounded shape of the sore were almost always present. The introduction of disinfectants into our military hospitals has done much to prevent the prevalence of this disease, which committed such ravages during the Peninsular war.

The French suffered most dreadfully from hospital gangrene in its worst form. The system they pursued of removing their wounded and operated cases from the camp to Constantinople at a very early date, the pernicious character of the transit, the crowding of their ships and hospitals, all tended to produce the disease, and to render it fatal when produced. Many of their cases commenced in camp, but the majority arose in the hospitals on the Bosphorus, where the disease raged rampant. In the hospitals of the south of France it also prevailed, and, from what M. Lallour, surgeon to the "Euphrate" transport, tells us in his paper on the subject, it must have committed great ravages in their ships, from one of which, he says, sixty bodies were thrown over during the short passage of thirty-eight hours to the Bosphorus. With them the disease was the true "contagious gangrene," and attacked, not only open wounds, but cicatrices, and almost every stump in their hospitals. They employed the actual cautery, after the manner of Delpech and Pouteau, with apparent success, to arrest it. The perchlorate of iron, charcoal, the tincture of iodine, lemon juice, &c., they employed as adjuvants.

In both the French and Russian hospitals, gangrene was often combined with typhus, and in such cases the mortality was fearful.

In the Crimea, during the heat of the summer of 1855, after the taking of the Quarries, and the assault on the great Redan in June, not a few amputations of the thigh were lost, from moist gangrene of a most rapid and fatal form. In the case of a few who lived long enough for the full development of the disease, gangrene in its most marked features became established, but most of the men expired previous to any sphacelus of the part—overwhelmed by the violent poison which seemed to pervade and destroy the whole economy. This form of the disease occurred in four cases under my own charge, in men who had had a limb utterly destroyed by round shot or grape. In all the knee joints were crushed, the collapse was deep and prolonged, and the operation performed primarily in the middle third of the thigh. Three of the four were of very intemperate habits. All these cases took place about the same time, at midsummer, when many other similar cases appeared in camp. The wards, though full, were not overcrowded, and could, from their construction, be freely ventilated. The weather was sultry, and cholera was in the camp. The atmosphere was surcharged with electricity, and the dreaded sirocco prevailed. Wounds generally assumed an unhealthy aspect for days, when this pestilential wind blew. The cases of all those who died in my wards seemed to be doing perfectly well up to sixteen hours, at the furthest, before death. Three of them were seized on the eighth day after amputation, just as suppuration was being established. The fourth died on the fifth day. The seizure and consequent symptoms were identical in them all. In recording one case I relate all. During the night previous to death, the patient was restless, but did not complain of any particular uneasiness. At the morning visit, the expression seemed unaccountably anxious, and the pulse was slightly raised. The skin was moist, and the tongue clean. By this time the stump felt, as the patient expressed it, heavy like lead, and a burning, stinging pain had begun to shoot through it. On removing the dressings, the stump was found slightly swollen and hard, and the discharge had become thin, gleety, coloured with blood, and having masses of matter like gruel occasionally mixed with it. A few hours afterwards, the limb would be greatly swollen, the skin tense and white, and marked along its surface by prominent blue veins. The cut edges of the stump looked like pork. Acute pain was felt. The constitution, by this time, had begun to sympathize. A cold sweat covered the body, the stomach was irritable, and the pulse weak and frequent. The respiration became short and hurried, giving evidence of the great oppression of which the patient so much complained. The heart's action gradually and surely got weaker, till from fourteen to sixteen hours from the first bad symptom, death relieved his sufferings. All local and constitutional remedies which could be thought of were equally powerless—nothing could relieve the system from the weight which seemed to crush it, or enable it to support the severe burden. Strong stimulants were the only remedies which appeared to retard the issue for a moment. Post mortem examination, instituted shortly after death, showed the tissues of the limbs, and in many cases those of the internal organs also, to be filled with gas, and loaded with serous fluid. The vessels leading from the stump were healthy, and in only one case had there been any actual mortification previous to death. The intestines, in two of the four cases, were much diseased. Was the cause which gave rise to this affection referable to "weakness or defective powers of action," arising from the patients' bad state of general heath, or "excessive irritability or disposition to act," from their being of intemperate habits? or was it "excessive irritation or excitement to act," arising from the severity of the injury sustained? After the taking of the city in September, the same form of disease again appeared, especially among the Russians who had been operated on; and was so deadly, that in no case, which I could hear of, did recovery follow.*

Erysipelas was latterly rarely seen in our hospitals. Several cases which appeared in my own wards readily yielded to treatment. At Scutari there were a good many cases, at the time when the men were most depressed by their hardships; but it was seldom virulent.

The troops suffered greatly during the first winter which they passed in the Crimea, from *frost-bite.* Death not unfrequently followed on the injuries it occasioned. The severity which marked these lesions did not arise from the degree of cold, as the temperature was never so low as of itself to cause the severe results produced, but rather from the depressed vital power of the soldiers, who could not resist the effects of a degree of cold which would have little injured them if they had been in rude health. The practice, which was nearly universal, of sleeping in their wet boots, aided greatly in causing the results. This custom arose from the fear, that if the boots were put off, they could not be drawn on again. They were retained, and thus the feet kept for a long time at a low temperature, with the circulation retard-

---

* Dr. Taylor, in his report on the 29th regiment, to which interesting document reference has been already made, says: "It is to be observed, as illustrating the possibility of gangrene infection lying dormant for some days, or of a fomites of the disease hanging about the clothing of the men, that wounded men discharged fit to rejoin their regiments, were, in several instances, returned from camp to hospital with hospital gangrene."

ed, at length lost their vitality—slowly, but all the more surely on that account.

The scorbutic poison, too, with which the men were drenched, predisposed strongly to the action of the cold, and it was even at times difficult to say how much of the destructive result was due to the one cause or the other. During the first winter the frost-bites were much more severe than during the second, and much more difficult to manage, from the more depressed vitality of the patients. I referred in a previous chapter to that peculiar effect caused in the feet by the union of scurvy and frost-bite, to which it is so difficult to give a name.

Tetanic symptoms resulted in a few cases from frost-bite injuries of the feet. The French suffered more than we did. In their hospitals a limb might be seen sphacelated half way to the knee. Uncontrollable diarrhœa was a common complication in such cases, and invariably, according to M. Legoust, caused death. Scrive (Mem. de Med. et de Chir. Milit., vol. xvii.) tells us that on the 21st January, 1855, with the thermometer at 5°, they had 2,500 cases of frost-bite admitted into their ambulances, of whom 800 died, and that at that period no operation succeeded, so that "it was necessary to abstain from operating." M. Legoust says he found, in treating his cases at Constantinople, that a solution of sulphate of iron formed the best dressing, but of its use I had no experience. To obtain the separation of the scars, and regulate the subsequent granulation on general principles, was what had chiefly to be attended to. Soothing applications appeared to be the best in the cases which I had an opportunity of watching.

It is not easy to decide whether or not we should operate in such severe cases as sometimes occur, when half the foot, for example, or the lower part of the leg is implicated. Either step is somewhat hopeless; but if the part be unquestionably dead, and of such a nature and size as not to be separable by the "parsimonious industry of nature," without so long a period of irritation and suppuration as will be, in all probability, fatal; or, if the presence of a large gangrenous surface endangers not only the patient himself, but also his neighbours in the ward; or if there is hospital gangrene present in the hospital, to an attack of which the long open state of his wound will so much expose him, then it is a fair question to consider, whether amputation is not a lesser evil than waiting. The success which follows, will depend much on the state of the patient's general health, and on the condition of the parts; as, unless a clear line of separation be formed, and the parts above be tolerably healthy, the irritation occasioned by removal will be sure to cause gangrene in the stump—at least so it was in the Eastern hospitals, in all the cases in which I knew it tried. I never heard of any amputation performed under the above circumstances succeed during the first winter, but several such occurred during the second. Operating at some distance beyond the spread of the disease, was generally found safer than at the place of division, between the dead and living parts.

Any wounds from frost-bite are peculiarly difficult to heal. Many suffered from their effects for months after getting to France or England.

The removal of bone from the toes or fingers, however black and apparently dead, and though only attached by the most slender connection, was certain to cause a great amount of irritation, which sometimes became most alarming. This result was probably as much due to the enfeebled state of the patient, as to the cause for which the operation was performed. Complete non-interference during every stage of treatment, the use of the mildest dressings, the removal of parts only when quite disjointed, proved the best line of procedure. I never saw any other followed, either in our hospitals or in those of the French, without there being ample cause to regret it. Any roughness even, in dressing these injuries, endangered the appearance of gangrene, on the verge of which they always seemed to hover.

CHAPTER IV.

INJURIES OF THE HEAD.

From April 1, 1855, to the end of the war,* the returns show a total of 630 cases of gun-shot wounds of the head attended by contusion merely, more or less severe, and 8 deaths are recorded among these cases. Of gun-shot fracture without *known* depression, 61 cases appear, and 23 deaths therefrom. Of cases of fracture and depression, followed by sensorial disturbance, 74 cases are mentioned, and 53 deaths therefrom; while of wounds penetrating the cranium, 67 cases and 67 deaths are recorded. Of 19 cases in which the skull was perforated, all died. The trephine was employed 28 times, and of these cases 24 ended fatally.†

Mr. Guthrie has said, with much truth, that, "injuries of the head affecting the brain are difficult of distinction, doubtful in character, treacherous in their course, and, for the most part, fatal in their results." Of all the accidents met with in field practice, these are, beyond doubt, the most serious, both directly and remotely—the most confusing in their manifestations, and least determined in their treatment, although they have engaged the attention of the master-minds of all ages and countries, from the time of the old surgeon of Cos, down to the present day. Such men as Petit, Quesney, Ledran, Pott, Dease, Heister, Cooper, Dupuytren, Bell, Velpeau, Larrey, Brodie, and a host of other honoured names, have thrown the light of their large experience and commanding genius on the subject; even minor points connected with it have been made the theme of whole libraries, and of innumerable discussions in the first medical societies of the world; still, there is no accident which the surgeon takes charge of with more fear and hesitation, as in no class of cases does he feel so much the mystery which surrounds

---

* The returns are not complete before the date specified.

† Alcock reports 28 cases of fracture of the skull from gun-shot, and 22 deaths. Meniere gives 10 penetrating wounds by balls, all of whom died—half on the day of admission. In the medical reports from India, I find only 9 cases so detailed as to be useful. They were all penetrating wounds, and 6 of them died. Lenté, in his statistics of the New York hospital, mentions 128 cases of fracture of the skull, attended by death in 106 instances. Several of these were fractures of the base, and none by gun-shot.

and guards our life: for while in some cases death follows the most trivial injury, in others, a vast amount of destruction, and even removal of brain-matter, causes little, if any disturbance.

In war, injuries of the head of all descriptions are presented to us. Those by *contre-coup*, especially such as implicate the base of the skull, are certainly rare; but these also at times do occur. The comparative rarity of this form of injury in military, as compared with civil practice, is possibly accounted for by the less frequent occurrence of such accidents as are fitted to injure the base, and by the fact that war-projectiles seldom present a surface so large as to supply those conditions, which the experiments of Bichat would show are necessary to produce fracture of the skull by counter-stroke. It is, however, by no means true, that "the punctured fracture," as it is termed, is the only species of injury to which soldiers in the field are liable. Shell, grape, and sword wounds of the skull, afford examples of almost every kind of fracture.

The nature of the injury, inflicted by a ball striking the skull, will depend chiefly on the angle of incidence, and the velocity. The character of the ball, too, has more to do with the matter than is generally supposed. If the direction of the projectile be very oblique to the surface, and if the force be exhausted at the moment of contact, then the injury may be very slight — a mere contusion of the soft parts or of the bone.* If the force be greater, than the pericranium may be much injured, the bone considerably bruised, or slightly fractured throughout its whole thickness, or in one or other of its tables separately,—the fracture of the inner sometimes taking place without any apparent injury of the outer. Further, the brain may be injured as well as its case, when the blow is yet more direct or severe. This injury may be merely of such a nature as, John Bell well says, "we choose to express our ignorance of by calling it a concussion," which may pass away, doing little harm, or which may be followed, at an uncertain interval, by encephalic inflammation, and compression from effusion.

Again, the effect of a ball "brushing" over the skull may be such, that while the bone is not fractured, the vessels between the skull and the dura mater may be ruptured, or the longitudinal sinus may be opened, as occurred in one case in the Crimea, and which has been related by the surgeon of the 19th regiment, in which it was observed.

A remarkable instance, showing how completely the skull may

---

* Stromeyer supposes that the danger of a grazing shot arises very much from pyæmia. Inflammation of the bone follows the injury, the veins of the diploe become implicated, and thus pus enters the system.

be destroyed by a glancing shot, without the scalp being implicated, occurred at the Alma. A round shot "en ricochet," struck the scale from an officer's shoulder, and merely grazed his head as it ascended. Death was instantaneous. The scalp was found to be almost uninjured; but so completely smashed was the skull, that its fragments rattled within the scalp as if loose in a bag. The condition of the brain was, unfortunately, not examined.

A bullet, from the great force with which it impinges upon the skull, and the concentration of that force on a small point, causes a fracture dissimilar to most of those which are met with in civil practice. It is this concentration of the force on a small point which renders fractures from a ball so dangerous, as the bone is driven deeply into the brain, and the splintering, especially of the inner table, is often very severe.

The greater splintering of the inner than of the outer table, by a ball penetrating the skull from without, is explicable on the principles which interpret the difference between the wounds of entrance and of exit in the soft parts, and which I before explained by reference to a series of experiments on planks of wood. The greater support afforded to the outer than to the inner table, by the parts lying behind it, and the diminished force of the ball as it passes through each, sufficiently account for the difference. An observation of Erichsen's on the point, quite supports this explanation. He has noticed that the characters of the apertures in the two tables were reversed, in an instance in which a man had committed suicide by shooting himself through the head from the mouth—the ball thus passing from within outwards. In a case from Bagieu, related by Sebatier, the same circumstance is noted in a similar instance.* The preparation in the Fort Pitt museum numbered 2592, illustrates the same thing. In that case, a ball had perforated the head, thus making two holes, the one in the front, and the other in the lateral and posterior part. The inner table of the orifice of exit is regular, while the outer "is torn up to an extent much larger than the ball." An appreciation of these distinctions is of much use to the medical jurist.

The character of the fracture caused in the skull by the large conical balls is, I am inclined to believe, considerably different

---

* Larrey thinks that in young persons a ball may enter the skull, leaving a hole less than itself, from the yielding and subsequent closure of the osseous fibres. This is not observed in the old, in whom the bone is more brittle, and splinters. A case is related by Dr. Longmore, of the 19th regiment, in the second volume of the *Lancet* for 1855, by which it would appear, that a ball may split, part enter the skull, and yet the bone recover its level by its resiliency so completely as to leave no trace of the passage of the part of the ball which entered.

from that occasioned by the round ball. The destruction by them, of the outer table, always appeared to me much greater than by the round ball; and thus, perhaps, it is that the size of the openings in the two tables is more equalized in the wounds occasioned by the former, than by the latter species of missile. So it comes, I think, that the true "punctured fracture" is less seen now in military practice than it was formerly. I state this, however, with much hesitation; as it would require a larger number of observations than I possess to substantiate it.

Balls striking the head otherwise than perpendicularly to its surface, or impinging against one of its angles, may be split,—part entering the skull, and part flying off. This occurred in cases which have been related by Mr. Wall, of the 38th regiment, and by Dr. Longmore, of the 19th. Such instances are not uncommon in war. Larrey, following the half which entered, removed it by counter opening from the back part of the head. One half of a split ball has been seen to lodge between the tables of the skull. The whole ball, also, has been found thus placed, especially at the fore part of the head. There are various instances on record of a round ball penetrating the outer paries of the frontal sinus, without injuring the inner table; but I believe that no such instances will ever be found where a conical ball is used. It not only penetrates, but generally perforates the skull, and almost always proves fatal.

The most dreadful injuries of the head seen in war are those occasioned by shell. Although rarely, yet it does at times happen, that this missile cuts open the scalp only, or merely grazes the bone; yet it more frequently occurs that large masses of the skull are driven by it into the brain. Examples will be afterwards given of shell wounds of the head. One of the most ghastly injuries of the skull which I ever witnessed, was caused by a fragment of shell. The whole frontal bone was driven deeply into the brain, yet, strange to say, the poor sufferer lived for twenty-four hours after such a wound.

Sword-cuts sometimes, as is well known, slice away parts of the skull. These portions will, at times, re-adhere, if immediately applied. In the museum of the Val de Grace several remarkable examples of this are to be seen. I had under my charge, after the fall of Sebastopol, a Russian soldier who had received such a wound, although the bone was not entirely detached in his case. The left parietal bone was cleft, so as to be almost separated. He would allow no one to touch his wound, except a comrade. His recovery was complete, the brain never showing any tendency to protrude, although quite visible throughout the whole extent of the wound. I saw this Russian in the interior, after peace, in perfect health. The comparative rarity of hernia of the cerebral substance after sword, as compared with gun-shot wounds, is very remarkable.

Cuts from a blunt sword are peculiarly dangerous, from the

extensive splintering and depression of the inner table which so commonly results.*

One of the most remarkable circumstances connected with gun-shot wounds of the head is, that they are not more universally followed by concussion, or that the symptoms of concussion, when produced, are often so temporary in duration. I have been frequently told by men who had received wounds of considerable severity, that they experienced merely feelings of passing "weakness" when struck. Symptoms of concussion, however, more generally follow severe blows, and the gradual, and almost insensible manner in which this state passes into one of compression or of inflammation, and that into consecutive compression, forms one of the most treacherous and dangerous features of these cases. It is evidently a matter of much importance to

---

* That trephining does little good in these cases, is illustrated by the practice of Dease, who had under his charge many men wounded by the "hanger," which played so important a part in all the street frays of his time. Four of the seven cases he trephined died; while in the only four instances in which he seems not to have interfered, recovery followed. In a case which, although not caused by a sword, was yet a fracture of a similar description, and which occurred lately in the Royal Infirmary of this city, (Glasgow,) under Mr. Lyon, the recovery was probably owing to the non-interference with the injured bone, further than the removal of loose portions. A man aged 19 was admitted on the 28th of July. He had been struck on the head by the handle of a crane, and the whole scalp round and round the head, with the exception of the anterior part, separated. The bone was fractured into small pieces, to the extent of four inches by one and a half, over the right side of the head—the fracture slanting obliquely over the orbit. He suffered much from the shock when admitted, but replied to questions put to him. Bony spiculæ were driven into the right eye, the right malar bone was broken, and the frontal sinuses opened. The loose and broken bones were removed, when the brain was found to be laid bare to the extent of three inches, and the dura mater destroyed. Low diet, purging, and cold locally, were the remedies—the scalp being carefully laid down, and the spiculæ removed from the eye. The fractured bones were not interfered with, further than the removal of perfectly loose portions. A week after admission the brain began to protrude by the opening in the skull, but by gentle compression it soon receded, and the patient made a rapid recovery, interrupted only by a slight hæmorrhage from a vessel in the scalp, which was easily suppressed. The wound completely healed—the bone being bridged over by dense tissue, and the cicatrix sunk in a narrow furrow, the pulsations of the brain remaining visible. His pulse never exceeded eighty. The supra orbital ridge remains much below its proper position, and the right eye is destroyed.

those who advocate trephining in certain circumstances, to be able to distinguish accurately between these variable conditions; as to operate in cases of mere concussion, or in a state of inflammation, would be murder—yet, how to discriminate, is a practical puzzle in many cases—especially in a large number which fall to be treated in the field, when the period of their coming under observation is very uncertain, and when no account can be got of their history, or early symptoms. It requires but the most cursory reading of surgical works to determine, that the utmost confusion has always existed between these various pathological conditions; even Sir Astley Cooper, with all his habitual clearness, has not unfrequently confounded them. It is little wonder that it should be so, as their clear distinction is found only in books, and their interdependence and mutual re-actions, as well as the uncertainty of their respective manifestations, all contribute to deceive "the pride of our penetration," and lead us into error.

The absence of any ascertainable cause, and the threatening symptoms which were present in the following case of concussion, interested me a good deal at the time. In former days it would have been infallibly ascribed to the wind of a ball. Quin, a private in the 18th Royal Irish, suddenly fell down unconscious, in the advance on the Redan, early in the morning of the 18th of June. He never could tell how this happened, not being aware of any injury. He was brought into my ward insensible a few hours afterwards. His symptoms were those of severe concussion. The surface of his body was cold, his respiration was slow and regular, and his pupils were contracted. No injury could either then or afterwards be discovered. Warmth, and an enema of the arom. sp. of ammonia, helped to restore him to consciousness, after he had vomited. He continued, however, for some hours, like a man half-drunk. Re-action was so violent as to call for bleeding, cold to the head, antimonials, and purging, to moderate it. Some days afterwards he suddenly became delirious, with injected eyes, one pupil being contracted, and the other a little dilated. He complained much of his head, which he afterwards said had felt all the time as if strongly bound by a cord. There was never any paralysis, or subsequent unconsciousness. By free purging, shaving the head, applying cold, restricting him to very meagre diet, and latterly, by the use of blisters to the nape of the neck, he completely recovered, though for about a month he suffered from severe headache, double vision, and a pulse unusually slow, and little changed on assuming the erect posture.

The danger occasioned by gun-shot wounds of the head will depend much on the part struck. At some places the ball is more apt to glance off than at others, while the strong processes of bone, the situation of blood-vessels, and the apparently greater necessity to life of some parts of the brain than others, introduce many elements into the calculation of the result. Notwithstanding all this, however, the curious eccentricities which characterize these injuries—the slight disturbance created by some, which, to all appearance and experience, are ten times more severe than others that prove fatal, upset our preconceived opinions; and while they puzzle us to account for the difference, they prove the truth of Liston's aphorism, that "no injury of the head is too slight to be despised, or too severe to be despaired of."

Generally speaking, it appears tolerably certain that wounds of the side of the head, especially anterior to the ear, are the most dangerous to life; and that a descending scale will give the following order—the fore-part, the vertex, and the upper part of the occipital region; the last being decidedly the least dangerous. Remarkable exceptions to this graduating scale of danger do, however, occur.

There are, at the same time, other circumstances besides the seat and nature of the injury which influence the result. The age of the patient, is, perhaps, the most important of these. With children and young persons, the same gravity by no means attaches to the prognosis of head injuries, as to similar accidents occurring to the old. Mr. Guthrie has well observed, that in the accounts of wonderful escapes and successful operations on the head, the subjects have been, in general, below puberty. The temperament of the patient, his excitability, his social condition, as giving rise to more or less anxiety regarding the result of his case—the means there are of carrying out his treatment as to quiet, isolation, &c.—the place where he is treated, whether in the hospital of a populous city, where the results of such cases are usually so fatal, or in the country, where so much more can be accomplished,—all these are important items in forming an opinion regarding injuries of the cranium.

Gun-shot wounds of the head, being chiefly compound, enable us to ascertain, with tolerable precision, the amount of injury which has been inflicted; and if it be thought necessary to employ any means to elevate depressed bone, we can do so with less hesitation than if the scalp was unhurt; as, if it be true what some of our best surgeons tell us, that the danger of inflammation in the membranes is increased by opening the integuments, then this source of danger cannot be charged to us. Such facilities should not, however, make us less careful in our proceedings.

As to the use of the trephine—the cases, and-time for its application—less difference of opinion, I believe, exists among the experienced army surgeons than among civilians; and I think the decided tendency among them is to endorse the modern "treatment by expectancy,". and to avoid operation, except in rare cases. In this, I believe, they judge wisely; for, when we examine the question carefully, we find that there is not one single indication for having recourse to operation, which cannot, by the adduction of pertinent cases, be shown to be often fallacious; while, if we turn to authorities for advice, we find that not a great name can be ranged on one side, which cannot be balanced by as illustrious on the other.

Simple contusion, without fracture or depression, caused the old surgeons to "set on the large crown" of a trephine, in order to prevent future danger. Fracture, although not accompanied by depression, or any other untoward symptom, called for the trephine in the practice of the Pott school; while many, even now, would operate to cure the local pain which so often remains persistent at the place of injury. Other surgeons, again discarding and condemning all this, say we should trephine only when there is depression; but the amount of depression which demands it, each interprets according to his own fancy. None knows so well as the army surgeon how very considerable a depression may exist, especially at some parts of the head, without any injury to the brain; nor how innumerable are the cases in which great depression has been present, without causing harm at any subsequent period of the patient's life.

A musket ball being the wounding cause, would appear to some a sufficient reason why the trephine should be applied, however slight may be the lesion. "We should always trephine," says Quesney, "in wounds of the head caused by firearms, although the skull be not fractured." "All the best practitioners," says Pott, "have always agreed in acknowledging the necessity of perforating the skull in the case of a severe stroke made on it by gun-shot, upon the appearance of any threatening symptom, even though the bone should not be broken; and very good practice it is." Boyer and Percy are equally urgent when a ball has caused the injury. However, "the experience of war," to which Quesney appeals in confirmation of his opinion, now-a-days completely condemns the practice, whatever it may have done formerly.

Further, "symptoms of compression" setting in early or late, are laid down by others as urgently demanding the removal of the bone. "No injury," says John Bell, "requires operation except compression of the brain, which may arise either from extravasated blood, or from depressed bone, or matter generated within the skull." But, unfortunately, we can seldom diagnose the existence of compression with any amount of certainty, when it sets in early, and experience teaches us that each and all of those signs which are said to indicate it may, under appropriate treatment, pass away without interference; especially when these symptoms appear early, and often, also, when they set in late. Compression, too, when it appears at a late date, if it arise, as it generally does, from the presence of pus, is well known to be seldom relieved by trephining. Dease first showed how it was that the matter was commonly deeply placed or diffused in such cases; and the instances in which it has been found on the surface, or evacuable by such a bold manœuvre as the well-known thrust of Dupuytren, are exceedingly rare.

Some authors, again, would have us trephine only when the symptoms of compression are severe, go on increasing in severity, and have continued for some time; yet, even under such circumstances, "recovery not seldom disappoints our fears, and mortifies us by our success."*

But, finally, it is to those surgeons who instruct us to operate when certain pathological conditions exist, which they carefully define, but which experience, unfortunately, tells us do not often manifest themselves by any recognizable signs, that we are chiefly indebted for useful directions to assist us in cases of difficulty. What good can it do to say, you must trephine when the internal table is splintered more extensively than the external, when effusion has taken place on the brain, and so on, when we have often no means of knowing when these conditions exist, or when we are fully aware that they have, each and all, been present, and that to a very considerable extent, without any of their appropriate signs being manifest?

But to refer more particularly to those cases which fall to the charge of the military surgeon. There are three classes to which the trephine is still occasionally applied: 1st, fracture with depression, before symptoms have appeared; 2nd, fracture with depression, attended immediately with signs said to indicate compression; and 3rd, fracture, with or without depression, followed at a late period by symptoms evidencing compression.

It is with reference to the first class of cases that "the experience of war" is most useful, and most decided. There are, I believe, very few surgeons of experience in the army now-a-days who approve of "preventive trephining." † It may be said in

---

* See, especially, as good instances of this, Quesney's first and second observations. In the first, the stupor and delirium lasted three months, and in the other, it had continued also for a lengthened time. Stromeyer, by antiphlogistic remedies alone, saved several in which the "stupor had lasted for weeks together."

† "That blood may be effused," says Guthrie, "and matter may be formed is indisputable, even under the most active treatment; but that any operation by the trephine will anticipate and prevent these evils, cannot be conceded in the present state of our knowledge; and the rule of practice is at present decided, that no such operation should be done until symptoms supervene, distinctly announcing that compression or irritation of the brain has taken place. It is argued that, when these symptoms do occur, it will be too late to have recourse to the operation with success; this may be true, as such cases must always be very dangerous, but it does not follow, and it never has been, nor, indeed, can it be shown, that the same mischief would not have taken place if the operation had been performed early."

our time to be a practice of the past—a practice to be pointed at as a milestone which we have left behind. A very large number of instances fell under my own notice in the East, in which, by the use of evacuants and quiet, and the absence of all operative interference, a perfect and uninterrupted recovery followed these injuries, even when the bone was very extensively depressed. Every surgeon in the army can recount many such cases. *If* any patients were lost from not having been operated on, I never saw any of them; but I do know of some patients who died, because they were subjected to operation.

The wonderful manner in which the brain accommodates itself to pressure, has been remarked in all times, and the crania in our museums show how extensive the depression may be, and yet the brain escape injury, or in which, although the central mass may be pressed upon or hurt, recovery has yet followed. In the cases of fracture with depression which have presented themselves to me during the war, the symptoms and the amount of depression have seldom been in correspondence.* But, in

---

* Hennen, in particular, refers to a case in which bone was depressed in "a funnel shape," to the extent of an inch and a half, and yet the patient lived in comfort for thirteen years. Stromeyer mentions forty-one cases of fracture with depression from gun-shot, and in many of which it is probable that the brain was injured, although that could not be ascertained. Of these cases only seven died, and one of these perished by typhus fever. All the rest recovered, and in only one case was there any operative interference, although signs of secondary compression appeared in several. The antiphlogistic treatment, carefully carried out, was alone adhered to. Seutin, who was at the head of the medical service at Antwerp when it was besieged in 1832, gives us the results of his experience in the following words: "Far be it from us the pretension to decide the question which divides practitioners of the greatest merit; we will not take up the defence of either the one side or the other, but we think that it is necessary to limit to a small number the cases of fracture which demand the operation of trephining—an operation which often causes grave accidents, and the success of which is always very uncertain. The following facts, collected at the siege of Antwerp, prove, in an evident manner, that in the greater number of cases of fracture of the skull, when they are simple, or even comminuted, or with slight depression, we can often abstain from operating. It was by immediate incisions, and taking care to extract all underlying fragments, and employing mild dressings, and using antiphlogistics and revulsives, that we have been able to avoid the use of the trephine. It was by such methodic treatment that we have obtained such happy results in the case of the large number of wounded which have fallen under our charge."

The re-union of bone which has been depressed, with the rest of the skull, is well illustrated by preparations 2506, 2507, and 2512,

order to attain favourable results, it is absolutely necessary that great attention be paid to the management of the patient, of which I shall speak more afterwards.

Those who have read with attention the records of campaigns, must have often been struck with the numerous instances which are there recounted where men, with gun-shot depressed fractures of the skull, have recovered in circumstances which forbade any attention being paid to them. During hurried retreats and forced marches, this has often occurred. When privation was added to the absence of all surgical interference, these happy results were the more marked. In Larrey, Guthrie, Ballingall, and in the Indian reports, many illustrations of this are found. Dease, also, long ago recorded the observation, that "those patients who neglected all precepts, and lived as they pleased, just did as well as those who received the utmost attention;" at which we need not wonder, when we remember in what "the utmost attention" consisted. Thus it would seem as if severe fatigue, irregular, and it might be intemperate diet, are less injurious to men with fracture of the skull than the probings, pickings, and trephinings which form the more orthodox and approved practice. Deputy-inspector Taylor, in his able report on the wounded of the 29th in India, after referring to several wonderful recoveries from gun-shot depressed fracture of the skull, very appropriately remarks, that he attributed the fortunate results in these cases "to the system adopted of very cautious meddling with the wound." *

---

in the museum at Fort Pitt. In that numbered 2512, "part of the squamous portion of the temporal, and part of the parietal bone," is depressed three-quarters of an inch from the original level, and the diameter of the fracture is about three inches, yet the patient recovered perfectly, and lived as an officer's servant for three years, when he died of fever.

* I cannot deny myself the pleasure of recording a case which lately occurred in the practice of Dr. George Willis, of Baillieston, in the neighborhood of Glasgow, which is remarkable for the extent of the lesion, the period when the trephine was applied, and the perfect and rapid cure. William Donald, aged 36, a pit-sinker, a man of intemperate habits, but of strong frame, was struck on the 20th of June last, at 4 o'clock in the afternoon, on the left side of the head, by a piece of stone weighing thirty pounds, which had been thrown high into the air by the explosion of a mine he had constructed in the prosecution of his work. He immediately fell down insensible, and was put, in that condition, into a cart, and conveyed to his house, which lay two and a half miles from the place where he met with the accident. In about half an hour from the moment he was struck, and before he reached home, he slowly regained consciousness, and on his arrival at his own door

More difficulty exists as to the treatment of the second class of cases referred to before, viz : those in which there is fracture with depression, attended immediately by those signs which are usually said to indicate compression.

Compression is undoubtedly the evil against which the trephine is generally employed. But yet, with all that has been said on the subject, in books and lectures, I question whether we are sufficiently acquainted with the nature, seat,* or signs of com-

---

he was able to walk into the house with assistance. He was, however, unable to speak. Dr. Willis saw him about this time, and found a semilunar wound, about nine inches long, extending over the left side of the head, and curving over the ear. The flap of the scalp hung down over his ear, and a clot of blood covered the bone. On clearing away this mass of effused blood, the bone was found to be comminuted and depressed in an irregular crescentic shape, to the extent of four inches long by two broad. It was driven downwards to the depth of a quarter of an inch, and comprised part of the frontal and a portion of the parietal bones. The flap of the scalp was cleaned and replaced, and cold applied. Nothing else was done that evening. His pupils remained unaffected at all times, and his pulse never was much disturbed, but at the evening visit his mouth was found drawn to the left side. Next morning at ten o'clock, the speechlessness remained, but no new symptoms were added. The fractured bones were so firmly impacted that they could not be removed without the use of the trephine, which was accordingly applied at the upper part of the fracture, and when a piece of bone was thus removed, the rest were easily got at, and withdrawn. The dura mater was entire, and rose immediately in the wound. At each pulsation of the brain, blood flowed from between the skull and the membrane. Whenever the depressed fragments were removed, the tongue could be protruded, which before the operation it could not. It projected to one side. The speech did not return. The scalp was replaced and fixed ; he was purged, and put on low diet, and kept quiet, cool, and in the dark. By night he had again lost all power over his tongue, but recovered it next morning, and from that period his convalescence went on so rapidly, that in three weeks his wound had completely cicatrized ; he never had an uneasy feeling, and returned in perfect health to his work within six weeks of the period when he met with the accident. I saw him, by the courtesy of Dr. Willis, some time afterwards. He told me he never had had a headache since the day of his dismissal, although he acknowledged to have been repeatedly drunk. The cicatrix was firm, and considerably sunk, and the brain pulsations could be obscurely felt at one corner of the wound.

* I have myself known the trephine applied, in two cases, to injuries on the vertex of the head, when the compressing fracture existed at the base. Are we in cases of doubt to proceed as Heister directs? "Sometimes it is impossible," he says, "to discover the

pression, to warrant us in undertaking, at an early period at any rate, an operation of so serious a description, as all recorded experience has shown trephining to be, without more reliable and more clearly-defined evidence of its presence than is commonly thought to denote it. Symptoms which, by the dicta of books, were unquestionably those of compression, have passed off, in the experience of every one, under a treatment of which non-interference was the most important item; while in other cases such large quantities of fluid—blood and pus—have been found, post mortem, on the brain, as all recorded experience tells us *should* have caused a compression which yet never appeared. We find cases on record in which it is evident that traumatic encephalitis was mistaken for compression, and the skull trephined; and in some such instances good effects have followed, evidently from the local bleeding, which, in several of these cases, was considerable; or, perhaps, from the preliminary incising of the pericranium, which we know has, in some cases, succeeded of itself in removing symptoms analagous to those caused by compression.

Blood rapidly effused may cause early compression, which we know often passes off as the effusion is absorbed; or mere congestion, the result of injury, may give rise to the same symptoms, and be allayed by depletion; yet, if we trephine early, we may have only such conditions to contend with.

If the bone be very deeply depressed on the brain, and the patient be comatose, with stertorous breathing, slow pulse, and dilated pupil, then it may be admissible practice to use the elevator cautiously, with or without the assistance of Heys' saw; but in all cases in which the bone is not very deeply depressed, and in which these symptoms are not very decidedly marked, nor have continued for a considerable time, I do not believe any interference should be attempted.

It is too much the custom, I think, to deny or overlook the danger which arises from the operation itself. This is no place to inquire what is the source of this danger, whether it be the admission of atmospheric air to the membranes, as supposed by

---

particular part of the cranium which is injured, the patient in the meantime being afflicted with the most urgent and dangerous symptoms. In these cases it will be necessary to trepan first on the right side, then on the left side of the head, afterwards upon the forehead, and lastly upon the occiput, and so *all round* until you meet with the seat of the disorder." Even in recent times the same practice has been recommended by Benjamin Bell, who says we must "form the first perforation in the most inferior part of the cranium in which it can with any propriety be made, and proceed to perforate every accessible part of the skull till the cause of the compression is discovered."

Larrey and Stromeyer, or the renewed irritation and injury of the brain coverings, or, as others say, from pus poisoning; but the fact recurs that the most serious, and at times fatal symptoms, have followed the operation itself, in cases in which, contrary to expectation, the parts below the bone were found sound.*

Injury of the skull, followed at a late date by compression, is perhaps the most hopeless of all the circumstances in which the trephine can be used, yet it seems that in which it is most properly and incontestably employed. Rigors followed by vomiting, a rapid pulse, stupor, delirium and palsy, usher in a condition of things which, except in rare cases, is fatal. The longer the time which intervenes before the appearance of such symptoms, the more deadly does their indication appear to be.† It is well known

---

*The mortality which attends the operation of trephining needs little proof, as it is one of the best recognized surgical facts. Take such a statement as that of Stromeyer, who tells us that during the three years he attended the hospitals of Vienna, London and Paris, he had not met with a single successful case, while many severe injuries recovered which were left alone. In the New York hospital only one-fourth of their cases recovered, i. e., eleven cases out of forty-five. In ten of these the operation was prophylactic, and in thirty-two therapeutic; three of the former and eight of the latter recovered. In India I find a record of four cases of trephining for symptoms setting in late, and all ended fatally. In the Glasgow hospital register I find no record of a recovery after trephining. In University College hospital Mr. Erichson speaks of four cases of recovery in thirteen operated on, and in the Paris hospitals Nelaton tells that in fifteen years all their operations of this kind for traumatic effusion have ended fatally. Mr. Guthrie thinks the danger greater when the operation is performed late. He thinks the sooner it is undertaken, if it is to be had recourse to at all, the better, "believing the violence to be greater when done on parts already in a state of inflammation than when they are sound." Larrey expresses himself in almost the same words: "We say, then, that the trepan should be applied when it is decidedly indicated, before the invasion of inflammatory symptoms, which show themselves more or less promptly, according to the idiosyncrasy of the patient, his age, and the cause of the wound; and when it is developed, the operation should be delayed till these symptoms cease. If this second period does not present itself, it is better to abandon the patient, devoted to certain death, than to try a useless remedy which can only hasten his last moments."

† The late period at which dangerous symptoms may be set up, the total absence of any irritation caused by foreign bodies impacted in the brain, which is occasionally observed, are well shown in a case related by M. Manoury in his report on Roux's service during the year 1841. A student, with suicidal intent, shot himself by the mouth. The ball tore the jaw, but there were no head

that in the majority of these cases the pus is so situated that it cannot be evacuated by the trephine. It is either diffused over the brain, between its membranes, or collected in depots deep within its substance, or at parts distant from the seat of injury. In a considerable number of cases, however, it lies superficially, when its formation has been occasioned by a concentrated blow like that of a ball, and may be found collected beneath the place of injury. It is only in these latter instances that any good can be got from the use of the trephine; but such cases are sufficiently numerous in their occurrence to indicate its employment in all instances in which distinct signs of purulent collection set in at a late date. "It is plainly an abscess of the brain," says John Bell, "and as it is an abscess which cannot burst or relieve itself, though the trepan may fail to relieve the patient, yet without that help he will infallibly die." In this is expressed the true reason for its use in these most hopeless cases. It is, in fact, a last resource, which we are not justified in refusing to avail ourselves of.

Besides this, it is also true, that in a considerable number of cases in which the pus has not been found immediately beneath the seat of injury, it has been discovered, post mortem, but slightly removed from it, within the brain substance—so near that very little would have effected its evacuation; and it is also well known that success has followed the bold expedient, first practised by Dupuytren, of plunging a knife into the brain when the abscess was not found on its surface. The case will end fatally to a certainty, if the matter is not evacuated, and in the event of the attempt failing, such a step, if conducted with proper circumspection, will not add to the gravity of the case. The following case is mentioned, not only because of the late appearance of urgent symptoms, but also because of the position of the abscess found after death, which was situated as above referred to: A private in the 29th was hit by a ball above the eye. The frontal bone was smashed, and the ball was lost apparently in the brain. No head symptoms whatever followed. Some loose pieces of bone were removed, but two parts which were depressed were not interfered with. The antiphlogistic treatment was decidedly maintained. For three weeks no symptoms appeared to create alarm; at the end of that period, however, a good deal of local inflammation was set up, and the depressed portions of the bone, being found loose, were removed. Very little disturbance followed this step, and he was finally discharged, about four months after the receipt of the injury, appa-

---

symptoms. On the sixteenth day he was so well as to ask for his discharge from hospital, while on the eighteenth head symptoms set in, and rapid death ensued. The wad and the ball were found in the brain, and yet for a fortnight not the least sign appeared of irritation, or of the presence of such formidable bodies.

rently quite well. A month after dismissal he returned into hospital, complaining of feverishness, headache, and a hurried and excited manner. There was nothing particular found at the seat of injury. The cicatrix was in the same condition as when he left the hospital. The brain pulse was evident, as it had been since the bone was withdrawn. Coma occurred shortly after his admission, ending in death sixty hours from the first bad symptom. When the head was opened, the hiatus in the bone remained unchanged, only that the edges of the aperture were smoothed and beveled off, and somewhat darker in color than the rest of the calvarium. The dura mater was thickened, but entire, and adherent at the place of the wound. The other brain coverings were highly inflamed, and sero-purulent effusion existed between them. A small abscess was found in the substance of the brain, immediately below the place of injury, and behind this, but separated from it by a thin partition of cerebral substance, was a larger abscess in the anterior lobe of the brain, which communicated with the lateral ventricle of the left side. The small abscess had a distinct sac, but the larger one had not. Dr. Taylor, who reports the case, adds: "These collections of pus might have been of some standing, yet the patient had not a bad symptom up to sixty hours before death." It is very possible that dissipation after dismissal occasioned the sad and fatal result.

A soldier of the royal artillery was admitted into the general hospital on the 15th of November, on account of a shell wound dividing the scalp over the inner and anterior angle of the left parietal bone. He walked to the hospital, assisting a comrade who was more severely hurt than himself, and he complained so little that it was with difficulty he could be persuaded to go to bed. A piece of bone about the size of a shilling was found on examining his head, depressed to the extent of about an eighth of an inch at the seat of injury. He was purged, put on low diet, and his wound dressed simply. In five days he was allowed to rise and assist in the business of the ward, being put inadvertently, by the surgeon under whose care he was, on full diet and a gill of rum. No bad symptoms showed themselves for ten days. His bowels were permitted to get costive. His wound was nearly closed. On the morning of the fifteenth day from admission, he complained of giddiness, his pulse was rapid, and his face flushed. Leeches and cold were ordered to the head, and a purgative administered. He rapidly grew worse. The wound, now dry and unhealthy, gave out but a slight gleety discharge. He made many attempts to vomit, which was encouraged by an emetic. His pupils became widely dilated, but remained sensible to the action of light. A fortnight after the setting in of these symptoms, he was found to be hemiplegic on the *left* side. I saw him at this period for the first time. His respiration was sighing, and numbered twenty-two in the minute. His pulse was ninety, and contracted. His mouth and tongue were drawn to the *right* side. He was sensible when roused, but lay in a half state of sopor when not addressed. The next day the trephine was applied to the seat of injury, and the depressed bone removed or elevated. The dura mater was covered by a pulpy mass of lymph. No pus was found. Some spiculæ of the inner table which lay on the dura mater were withdrawn. His symptoms in no way improved. His tongue was next day drawn to the left side, but his mouth was unaffected. He had several severe convulsions over both sides of his body, and he died two days after being trephined. The skull was found fractured across the sagittal suture into both parietal bones. The dura mater was little detached round the seat of injury; but it was there dark and pulpy, having a semi-organized clot on its surface. The brain was softened at the place of injury, and had a clot as large as a walnut lying on it; while at two points on the opposite hemisphere, at the end of the longitudinal fissure, soft spots were found, about as large as a sixpence. Pus existed abundantly below the membranes, and bathed the surface of the right hemisphere, as well as extended to the base of the brain, between the hemispheres and under the cerebellum.

The neglect as to diet and the maintenance of the secretions, were probably the causes of death in the above case. It is certainly not always easy to maintain as careful a supervision on these points as is necessary, when no functional disturbance whatever is present, and the injury seemingly slight; but this is only one of the many examples which might be adduced to show the necessity of the long and careful watching which such cases require.

The above was one of the only two instances in which the trephine was employed in the general hospital, and both ended fatally. In the other case, it was used by one of my colleagues for signs of compression setting in early, with bone much and extensively depressed.

Finally, judging of this question from an examination of the writings of our great masters, the conclusion which presents itself is, that as the symptoms calling for the use of the trephine have been so variously interpreted by men of experience; that as the operation has failed as often as it has succeeded in removing the dangers apprehended; that as the good which has occasionally followed is ascribable, in many cases, to other concurrent circumstances, and not to the removal of the bone; and finally, that as the operation, *per se*, is not devoid of danger, we should never have recourse to the trephine unless the indications for its use are very decided, have been present for some considerable time, and have not been assuaged by other remedial measures.

Further, I am disposed, not only from reading, but also from the observation of not a few cases which fell under my notice during the late war, to conclude, regarding the cases and symptoms which demand operation—that *primarily*, operative interference (under which term is included the use of the trephine, saw, or elevator,) in gun-shot wounds of the head, should never be had recourse to except (1) in cases of fracture with great depression—cases in which the bone is forced deeply into the brain, especially if it is turned so that a point or an edge is driven into the cerebral mass; or (2) unless we clearly make out the impaction of spiculæ, balls, or other foreign bodies in the brain, which cannot be removed through the wound by means of the forceps; that *secondarily*, the cases which call for operation are (1) those in which a foreign body is at this period discovered irritating the brain, and which cannot be extracted without a piece of the bone being removed; or (2) those in which signs of compression set in after a well-marked rigor, continue to increase in intensity notwithstanding treatment, and have lasted for some time.

In the treatment of gun-shot injuries of the head, operative proceedings form the least important items, as they can commonly be avoided if the rest of the management be judicious, and their success will chiefly depend on a careful attention to less imposing, but more important measures.

In their examination the finger should alone be employed, and that even with much caution. They should not be enlarged, unless a more important object be held in view than to clear up doubtful points of diagnosis. If the bone be so extensively destroyed and depressed as to demand early interference, it will make itself sufficiently evident without its being necessary to incise the scalp for the purpose of making the distinction. Stromeyer fitly recommends the application of a piece of wet linen to the wound, which, as it adheres to the scalp, excludes the air. Cold—ice, if possible, or if it cannot be had, simple water—should be applied over this; the patient put to bed in a tent by himself; an active purgative administered, and a most meagre diet allowed. The utmost quiet should be enforced, and in short, the antiphlogistic treatment very decidedly and completely carried out. He should be visited frequently, and if any signs of inflammatory or excited action supervene, instant and copious bleeding should be put in force. "Of all the remedies in the power of art," says Pott, "for inflammations of membranous parts, there is none equal to phlebotomy, and if anything can particularly contribute to the prevention of the ills likely to follow severe contusions of the head, it is this kind of evacuation; but then it must be made use of in such a manner as to become truly a preventative—that is, *it must be made use of immediately and freely*." I never saw any good arise from the use of tartar emetic in these cases. Cold locally, purgatives, low diet, and early bleeding, repeated freely when signs of disturbance showed themselves; these, with the application of leeches in some cases to the head, seemed always sufficient, as they are the most useful means of treating such patients.

As to the extraction of balls when lodged in the brain, the rule, I believe, almost universally followed in the army, is to extract them if they can be at all got at. It is true that masses of a far more formidable nature than balls have remained on, and even in the brain without mischief, and that balls have been discovered encysted years after their entrance. But these cases form a mere fraction of the number in which the presence of the ball has determined fatal complications; yet they are the "ignes fatui" by which some would mislead us from the plain path of duty, which inculcates the removal of such foreign bodies, if at all practicable. Sir B. Brodie, arguing from an analysis of the published cases, advocates their abandonment unless superficially placed; but from this view nearly all military surgeons dissent. In our proceedings, however, "boldness must not partake of temerity." Few would have the courage or confidence of Larrey, or Sir Charles Bell, to follow and extract the ball from the side of the head opposite to the place of entrance, or, like Sedillot, pursue it to the depth of several inches in the cerebral substance; yet all reasonable attempts ought to be made for its extraction. "Nothing," says Sir George Ballingall, "will induce me to countenance the practice of leaving it there, except the impossibility of finding it;" and again, "I am of opinion that it ought to be extracted even at the risk of some additional injury; in short, the prohibition of violence ought rather to apply to the search after balls, than to the operation of extracting them." "We have already cited several cases," says Quesney, "which teach us that foreign bodies may remain a long time in the brain without causing death; but with this knowledge we must also bear in mind that it is our duty to extract these bodies, which, sooner or later, almost always prove fatal to the patients; and when we have reason to suspect from the events, from the instrument which inflicted the wound, or from the state of the fracture of the skull, that such bodies are retained and concealed in the substance of the brain, we should make the necessary examinations for the discovery."

If the ball has penetrated deeply into the brain, it is a matter of little moment what steps are taken. Perhaps the best line of conduct is to let the man die in peace. I have never known a case of perforating gun-shot wound of the head recover. Some such are, however, on record.

Cases in which pieces of loose bone remain on the dura mater, do not always require to be interfered with. Many surgeons of large experience in the Crimea, preferred leaving them to be thrown out by the natural effort, and were not particular even about keeping the wound open. However, I believe this practice to be often dangerous, and that loose portions of bone should always be cautiously removed. The evil effects of leaving them, as well as the injurious influence of too early a recurrence to a stimulant diet, were well marked in the following case: M'Louchlin, a private in the Connaught Rangers, aged 19, was admitted into the general hospital on the 8th of September. He had been knocked down, and rendered insensible, by a blow from a piece of shell in the final assault on the Redan. A scalp wound two and a half inches long, was found extending from before backwards over the vertex of the head, and a small piece of bone was observed to be depressed at its anterior extremity. The patient did not become conscious for twenty-four hours after admission. Purging, and low diet comprised his treatment. Cold dressing, and nothing else, was applied to the wound. He remained perfectly well, complaining only of slight headache and giddiness, for three months; small pieces of bone being discharged in the meantime from the wound, which had almost closed. After being about a month in hospital, he was allowed full diet, and a gill of grog daily. On the 8th of December, three months after receiving his wound, he complained of a sort of transient paralysis of the left arm, which, although it continued only for a second or two at a time, recurred frequently. His sense of smell, too, suddenly left him. There was no other symptom. On being questioned, he said he had had a rigor, and several "fainting fits" during the days immediately preceding that on which he first complained of the paralysis. Next day he had a more prolonged fit of paralysis during the night than he ever had had before, the attack being preceded by pain in the left side. I first saw him during an attack on the 9th of December, which was more severe and more prolonged than any preceding one. His left arm hung powerless, and there was complete anæsthesia of the left arm and side, from the clavicle to the false ribs, and from the line of the nipple to the spine. The left side of the neck behind the sterno-mastoid was also without sensation. His face was unaffected. The integuments around the wound were puffy, and very sensitive. He said that his uneasy feelings had gradually increased as the wound closed. His bowels were opened freely, and a light poultice was applied to the wound, which was incised. The fit he had on the 9th passed off, leaving the arm weak. The sensibility of the left side slowly returned during the succeeding days. The fits of paralysis came and went, his arm recovering its power, in a great measure, between them. A sharp bit of bone was at last observed lying on the dura mater, and when it was removed, the untoward symptoms disappeared. Shortly after this he came under my care. By quiet, and the use of unstimulating food and laxatives, he progressed most favourably; but on several occasions transient feelings of weakness—for there never again was a state of paralysis established—passed over the left side, when any scale of bone became loose, and lay on the dura mater, and, so soon as this was removed, these feelings left. If his bowels became costive, even for a very short time, not only did the headache and giddiness increase, but the numbness in the side returned. When he left for England no bits of bone could be discovered, and the wound was nearly closed; and he is now, I understand, doing duty with his regiment. Many of the symptoms in this case were those set down as calling for the use of the trephine; but the cautious removal of the fragments when loose, the local bleeding, and the purging, did all that was required.

Stromeyer warns us particularly against attempting too soon to remove pieces of necrosed bone, as he thinks they do little harm if allowed to remain. In this my own observation leads me by no means to agree. If the dead piece can be removed without violence, I believe it should always be done as soon as it is found to be loose.

On the treatment of hernia cerebi I have no remarks to offer.

Hardly less important than the immediate treatment of gun-shot wounds of the head, is their after-management. No class of cases requires more lengthened and careful supervision. Relapses may occur long after the patient is apparently beyond danger; and from the most insignificant causes—of which, perhaps, irregularities in food, the use of alcoholic stimuli, and retained evacuations, occupy the foreground—a chronic inflammatory condition of the membranes is apt to become established, which is no less difficult to manage, than dangerous in its ultimate results. Very many cases are on record in which men with balls embedded in the brain have apparently recovered completely, but have suddenly fallen down dead when they had got drunk or excited.

The following cases are added, as in some measure, illustrating injuries of various parts of the head. They are selected from a large number whose features are nearly parallel.

Hughes, an artilleryman, was admitted into the general hospital under my colleague, Mr. Rooke, on the 15th of November. He had been struck over the upper part of the occipital bone by a piece of shell, when the siege-train on the right attack exploded. He was rendered insensible by the blow. The scalp was considerably lacerated over the right upper part of the occiput, where a stellate fracture was found; part of the bone being depressed for about a quarter of an inch below the surface. He recovered some degree of consciousness a short time after receiving the blow, but was dull and stupid when admitted into hospital; answering questions if urgently put to him. His head was shaved, and cold applied. The next day he was rational; his eyes were bloodshot, but beyond this there was no bad symptom. Purging, and cold locally applied, were used. A few days afterwards he had headache, and intolerance of light. Dimness of vision, and flushing of the face followed, but there was no notable peculiarity in the pulse or pupil. Leeches were now applied to the mastoid processes; beyond this, the use of laxatives and low diet, nothing else was required to dissipate all threatening symptoms, and he left for England in January, quite recovered.

In the above case we had merely concussion at first, followed by a threatening of traumatic encephalitis. The treatment was simple, and the cure complete.

Clarke, private, 38th regiment, aged twenty-two, was wounded on the morning of the 18th of June, but was not brought into hospital till the evening of the 19th, as he lay where he could not be got at till the armistice. A piece of shell had struck him on the upper part of the occiput, laying the scalp open to an extent of two inches and a half. The bone, though denuded, was not seen to be fractured. His symptoms were dizziness, pain in the forehead, and great throbbing in the temples. He was quite rational, but dull, and had double vision and strabismus. His pupils were slightly contracted. His chief complaint then, and for some days after, was of his neck and lower jaw, which had received no injury; but the parotid and submaxillary glands were swollen on the wounded side—a symptom which I have observed in several similar cases. His pulse was forty per minute when lying down, and sixty-nine when he sat up. By active purging, and cupping the nape of the neck, and by the use of low diet, his bad symptoms gradually disappeared. For some days after admission his pulse did not change, except that on one occasion it fell to thirty-eight beats per minute; but as he got better, it rose to the healthy standard. On three different occasions, while he was under my charge, his bowels being unrelieved for a day, his bad symptoms returned in a modified degree, and his pulse sank; while, whenever his bowels were freely opened, all uneasiness vanished, and his pulse again rose. The alternation was most curious, and very rapidly developed. This case, like many others, illustrated well the marked sympathy which exists between the head and the bowels. The same slowness of the pulse was noticed by Dr. John Thompson, in the case of a similar injury after Waterloo.

A French soldier received a ball about an inch behind the left ear, which escaped above the eye of the same side. His antagonist, who shot him, was close to him at the moment he fired. This man fell down insensible, and was carried to the ambulance; but he recovered his senses before his arrival there. There was a little blood oozing from both openings; he was dull, but sensible, and complained much of a throbbing pain throughout his head. The ball having escaped, nothing was done for him, further than picking away some small loose fragments of bone, and applying wet dressing. He was freely purged, and got no food. In twenty-four hours the pain in the head having greatly increased, and being accompanied by delirium, with rapid pulse, ferrety eyes, and hot skin, he was largely bled, and cold was applied to his head. His symptoms were relieved, and from that day he never had a bad symptom: all the treatment his case required being merely low diet and free purging.

Another almost identical case occurred in our own hospital at Scutari, where I saw the patient under the charge of Staff-surgeon Menzies. The ball had in this case entered two inches behind the left ear, passed deeply, and was removed from the temple. Some hæmorrhage set in from both wounds, as well as from the ear, a few days after injury, but it was arrested by pressure. He was dull, and complained of headache for a few days after the occurrence of the bleeding; but by low diet and purging he made an excellent recovery, only that his hearing was destroyed on the wounded side.

A soldier, aged nineteen, belonging to the Second Division, was struck at Inkermann by a rifle ball, over the vertex of the head to the right of the centre line. The ball, passing from before backward, "furrowed" the bone, breaking both tables. This patient declared that he never lost his senses, but felt so weak that he had to sit down. He walked to the hospital, where he was twice bled, and actively purged. The bone along the line of the ball's passage being broken into small fragments, was removed with the forceps, and cold was applied. The brain was bared, but the dura mater, although scratched, was not found torn. A threatened attack of inflammation of the brain was successfully combated by repeated venesections and purging, and the patient made a good recovery; a sulcus about two inches long being felt by the finger over the vertex—the brain pulsations being distinguishable at one extremity of it.

An artilleryman was wounded on the 18th of June by a piece of shell over the back part of the head, and rendered insensible. He soon recovered, rose, and walked unassisted to the general hospital. No fracture was at first detected, and the lacerated scalp-wound which existed, was dressed simply by the surgeon under whose charge he fell. Headache alone was complained of for some days, during which period he was kept low, and freely purged. When the wound was nearly healed, he was unfortunately allowed butcher meat and a gill of rum. About a week afterwards, severe cerebral symptoms rapidly and suddenly showed themselves, and the wound took on an unhealthy action. The injury was now more carefully examined, the scalp being incised to assist the investigation. A fracture of the occipital bone was found. Bleeding was encouraged from the incision; leeches were placed on the mastoid processes; he was well purged, and cold applied to the head. His diet was again reduced. The unfavorable symptoms almost immediately subsided, and by the use of low diet and purgatives, soon totally disappeared, never to return. In this case a too generous diet doubtless caused the appearance of the unpleasant symptoms which supervened, and which, if not promptly arrested, would have been fatal. The local bleeding assisted materially; but the active purging, the cold applications, and the low diet were the chief means of saving him.

The following case, the particulars of which were kindly furnished me by acting Assistant Surgeon Brock of the 47th regiment, was a most interesting one, not only from the extent of the injury, but from "the phases of recovery."

Keefe, a private in the 47th regiment, aged 23, was struck, on the 15th November, by a piece of shell over the vertex of the head, and felled to the ground. When found, a short time afterwards, he was apparently dead. The surface of his body was cold, his pupils widely dilated and insensible to light, no respiration or motion of the blood perceptible. His face was much scratched and congested. Some blood flowed from the right nostril, and the superficial veins of his neck were gorged. The main wound in the scalp extended nearly from ear to ear, across the vertex of the head; and lesser wounds passed in different directions from this great one. The flaps of the scalp, formed by these wounds, were reflected in different directions. A large portion of the bone was seen to be destroyed, and the space left was filled by coagulated blood. The patient was seen by several surgeons, and so impressed were they that life was extinct, that he was carried to the tent set apart for the dead. Twenty five minutes afterwards, on being again visited, some faint signs of life were observed. There was a flutter at the wrist, and an occasional sigh. Profuse bleeding from the head followed, and on the clot, which was seen to be mixed with cerebral matter, being removed, it was found that the bones forming the vertex of the head were destroyed to the extent of $2\frac{1}{4}$ to $2\frac{1}{2}$ square inches. In this was included part of the superior angle of the occipital bone, and a part of both parietal bones, the sagittal suture being clearly defined along the centre of one detached piece. Part of this extent of bone was altogether gone, and the rest, being detached, was removed.

The surface of the dura mater was scratched, but not torn, except at one spot—at the lateral and posterior part of the wound—where it was lacerated, and from which a spicula of bone an inch long, and which was imbedded in the right hemisphere of the brain, was removed, a piece of cerebral matter the size of a nut adhering to it. The brain at this part seemed soft and broken down. Some depressed bone was elevated, and all loose scales removed. The scalp was brought together by suture, and lint wetted in cold water applied. Next day the patient was quite unconscious, lay on his back, and breathed regularly and naturally. His pulse was very weak, and his surface warm and moist. He passed his urine in bed. His pupils were dilated and insensible to light. He could swallow freely. During the two following days his state was unaltered. Both eyes became affected with strabismus. The treatment consisted of purging, cold to the head, and the most sparing diet. On the fifth day, there

were some signs of returning consciousness. He tried to change
his posture, and crossed his arms on his breast. His pupils, too,
acted feebly, and a profuse perspiration covered the surface of
his body. On the following day he again relapsed, and the
wound, which had begun to suppurate, now became glazed and
dry. When his bowels were got to act freely, he again improved
and became conscious. He complained of pain in the head and
down the left side of his body. Thus he went on till the eleventh
day, being conscious and able to speak. His bowels were carefully kept acting. His pupils had, up to this time, come to con-
tract and expand freely, and the wound was suppurating kindly.
He slept much, and expressed a great desire for food. On the
eleventh day he became suddenly restless and delirious, particularly at night. The strabismus returned. His eye became dull
and semi-glazed, and his pupils were widely dilated and little affected by light. By the eighteenth day these untoward symptoms had in a great measure abated. He was sensible and craved
for food. His left side was found to be paralysed, the face not,
however, being implicated. His pupils were still somewhat dilated, but active. There was also some œdema of the feet and
ankles. By the twenty-third day, granulations had formed round
the wound. Part of the scalp had adhered by the first intention.
His sleep was now natural and undisturbed. Except the temporary irritation caused by some spiculæ of bone, he went on improving from that time. Attention to his diet and the state of
his bowels, and allowing a free exit for the secreted pus, comprised all the treatment followed in this case. If his bowels were
for a day unrelieved, the bad symptoms immediately reappeared.
I examined him previous to his going to England; in January,
and at that period he was in every way recovered. The head
wound was entirely closed, but a depression to the extent of about
three-fourths of an inch existed over the site of the injury, and
the pulsations of the brain were quite perceptible.

I learn from Deputy-Inspector Taylor, that Keefe was invalided at Fort Pitt on the 28th May 1856, on account of "general
loss of sensibility and motion, partial in the upper, but most
complete in the lower extremities." He was in hospital at Chatham from 23d March to 26th June 1856, his state being as follows: "The wound on the head formed two sides of a triangle,
and is about two and a half inches in length on the right side,
and much longer on the left. It is quite healed, but there is a
very considerable depression. The pulsations of the brain are
quite perceptible. Complains of severe pain across the forehead,
of an intermittent type. Has lost the power of his lower extremities, with exception of being able to draw them up and
stretch them out in bed. Has not lost much flesh, and his general health and functions good." He thus appears to have relapsed after leaving the Crimea, as the marked paralysis he had
at Chatham did not exist when he left camp.

The intermittent headaches, spoken of in this case, are among the most troublesome sequences of injuries of the head. A careful regulation of the bowels and diet, with blisters to the nape, and morphia, appeared to me the best remedies. It is a remarkable feature in the progress of head cases, how often the setting up of sub-acute inflammation shows itself by an aggravation of the leading symptom—whatever that may be—which had existed before: the headache, palsy, or epileptiform fits. This was clearly defined in several cases.

The following is an example of a severe injury of the forepart of the head, caused by a piece of stone.

A French chasseur-a-pied was struck on the centre of the forehead, above the root of the nose, by a piece of stone about the size of a walnut, knocked up by a shell. The stone completely buried itself, and required some skill to extract it. Pieces of bone, comprising nearly the whole ethmoid, were discharged, and a large hole in the frontal bone resulted. Three days afterwards, transient, but easily-allayed head symptoms appeared, and he made a most excellent recovery, with a fistulous opening, however, remaining. The interest attaching to this case arose from the fact that the inner table of the skull was not fractured, and from the almost total absence of any head symptoms.

It is well known that balls may perforate the outer table of the skull on the forehead, without injuring the inner. Of this the above may be taken as an example; although a stone, and not a leaden ball, was the missile. Several cases occurred in the Crimea of another wound on the forehead which is curious, viz., such as are caused by balls passing from side to side of the head below the level of the brain, but destroying one or both eyes.

At Inkerman a French soldier was struck by a ball over the upper part of the left parietal bone. A comminuted fracture was caused; the bone to the extent of a square inch being so broken and detached, as to be removed at the first dressing. The dura mater was slightly injured, and a small spiculum, which had been driven into the brain, was withdrawn. He remained speechless for about a week, then articulated hesitatingly, and finally, about six weeks from the receipt of the wound, completely recovered his power of speech. The curious thing in this case was, that perfect anæsthesia of the thumb and two first fingers of the right hand existed from the moment of injury, without any loss of motion whatever, and that this slowly disappeared as the wound healed, and he recovered.

To multiply cases would be of little use. The teaching of all was to lead us to wait; to purge the patient thoroughly; to remove only such pieces of bone as could be got at with forceps, and which were quite detached and loose; to bleed, if need be, locally and even generally; to use cold applications when there was a fear of inflammation; to enjoin perfect rest, not only to the body generally, but, if possible, to give repose to the special senses also, by isolating the patient, and thus removing the stimuli to their exercise; to enforce the lowest diet, and to continue all this treatment for a long period, even after all danger seemed past; and, finally, to treat any incidental complications on general principles.

It is extremely difficult to get soldiers to avoid stimuli, or to attend to their secretions; and the desire for improved diet leads them sometimes to deceive one as to their feelings. The discipline of a field hospital can often be infringed, and as it is not easy to persuade men of the soldier's disposition, of a danger of which their sensations give no warning, it is necessary to watch them with great care.

Hepatic abscess I saw none of, and the nervous irritation and weakness, which so often follow injuries of the head, fell seldom under my notice, from the transference of the patients to the rear as soon as their wounds were healed. Jaundice was present in several fatal cases in which the head received injury.

CHAPTER V.

WOUNDS OF THE FACE AND CHEST.

After the 1st of April 1855, to the end of the war there occurred 382 cases of simple flesh contusions, and wounds of the face more or less severe, and one death is classed under this head. Of wounds penetrating, or perforating the bony structure of this region without injuring important organs, there were 107 cases, and 10 deaths; and of those accompanied by lesion of important organs, 44 cases appear, and 3 deaths. Most of the fatal results were owing to other concurrent causes.

Wounds of the face have been interesting chiefly from the rapidity with which even the most severe and dangerous-looking of them heal. The extreme vascularity of the tissues of the face endows them with a vitality which rectifies most injuries, and the surgeon is often enabled, both on this account and from their great distensibility, to repair the loss which has been sustained, even when that has been very extensive. It would be much easier to say where, and how the face has *not* been pierced by balls, than to enumerate the directions in which it has. The upper and lower jaws have been fractured, and large portions of them removed, yet, with few exceptions, a good recovery has followed, when no other concomitant injury assisted to bring about an unfavourable issue. One or other of the lower maxillæ, anterior to the masseters, has been carried away, and in one case which came under my notice, but which ended fatally, both lower maxillæ were removed by a round shot.* The upper jaw has been

---

* In a very interesting paper read to the Imperial Academy of Medicine, by M. Hutin, in April 1857, there is an account given of an inmate of the Invalides (to which M. Hutin is surgeon in chief,) who had the lower jaw carried away by a cannon ball at the battle at Wagram, forty-eight years ago. He recounts the changes which the parts have undergone since. It seems that the hæmorrhage was very severe at the moment of injury, but that it ceased spontaneously. The tongue hung down in front of the neck, and was never drawn into the throat—an accident which did not occur in four other cases, in which M. Hutin has known a little injury produced by a like cause. The patient referred to by M. Hutin has worn a silver mask since his accident, which protects his tongue hanging out, and adherent as it is to the neck. By means of this mask the variations of temperature do not affect the wide void

WOUNDS OF THE FACE.

completely destroyed, and in one case which occurred in the 31st regiment, a grape shot, seventeen oz. in weight, was impacted in the superior maxilla, and necessitated the removal of most of the bone.

Hæmorrhage is undoubtedly the great source of annoyance and danger in gun-shot wounds of the face. The difficulty of commanding it is at times so great as to place the patient in imminent danger. It frequently appears early, but stops spontaneously. Men who have received a severe face wound, seldom leave the field without sustaining a considerable loss of blood, and secondary hæmorrhage is common when the bones have been fractured. The depth, irregularity, and extreme vascularity of the parts make the application of a ligature to the bleeding points difficult, and to be effectual, compresses must be applied with much niceness. It is in wounds of the deep branches of the face, in which secondary hæmorrhage has taken place from a sloughing surface, that Anel's operation, performed on the main artery, may be said to supersede, from necessity, Bell's doctrine of local deligation.

The branches of the facial nerve are sometimes so much injured in wounds of the face, either by the ball, or by the fractured bone, that temporary and even permanent paralysis may ensue; but there is one source of danger in these cases which does not always obtain the attention its importance demands. I refer to the swallowing of the secretion from the wound. If great care be not taken to remove all the morbid secretion which results from injury of the bones of the face, if any amount of it gets into the stomach, much constitutional irritation will result, and a fever of a low typhoid, and very fatal form will be caused. I believe I have seen this result very clearly follow the cause referred to in some cases. In one case, where a sergeant of the Buffs died in the general hospital from the effects of a severe face injury, by which the anterior part of the lower jaw, and a small portion of the upper, were fractured by a round shot, I suspect the fatal result was at least accelerated by the cause mentioned, although the utmost care was taken to prevent its occurrence.

---

which exists in the floor of the mouth. The most remarkable change which the progress of time has brought about in the parts is, that the upper jaw, in place of preserving its horse-shoe shape, has become so contracted at its middle as to assume the figure of an hour-glass. This change began to take place three years after he was wounded, and has gone on increasing up to within a short time. The secretion and loss of saliva is great, but the patient enjoys perfect health. There is an interesting question raised by this case, viz., whether an analogous change may be looked for in those instances—of late years pretty numerous—in which the lower jaw has been excised.

He was a very unhealthy man, who had just been discharged from his regimental hospital a few days previous to the accident, and was of a nervous, irritable disposition. He was struck from the side by a small round shot, which had previously struck the parapet of the trench. The symphysis, and part of the body of the lower maxilla, as well as a small portion of the upper jaw, were destroyed. The soft parts, especially at the chin, were much torn and bruised, and ultimately sloughed. When examining his chest, on account of a cough which troubled him on admission, a cavity was discovered in one of his lungs. Hæmorrhage took place repeatedly from branches of arteries opened as the slough separated. By maintaining an opening below the chin, and washing the wound from the mouth, the greater part of the abundant secretion was removed; but yet, no small quantity found its way backwards into the throat, and was swallowed. His stomach became very irritable, his strength failed, and a low muttering delirium preceded death. A putrid abscess occupied the summit of one lung, and pus was infiltrated among the tissues covering the trachea.

In fractures of the bones of the face from gun-shot, we make an exception to the general rule of removing fragments which are nearly detached. The large supply of blood which is sent to every structure in this region, enables pieces of bone to resume their full connection with the other tissues, when detached, in a way that would be fatal to similarly placed portions in other parts. Hence the rule, not to extract any spiculæ whose attachment has not been completely destroyed, and whose direction is not opposed to a proper union of the broken parts. The exfoliation which follows in injuries of the bones of the face, is slight as compared with those of other parts.

The destruction, or injury, of one or other of the organs of special sense, and the deformity which may be caused, as well as the tedious exfoliations which at times follow severe face wounds, are the chief ulterior causes of suffering and annoyance to which they give rise. In cases in which the lower jaw is destroyed, the loss of bony substance, the powerful action of its muscles, which is so difficult to counteract, and the imperfect mode of repair, contribute to occasion a considerable amount of deformity. It is a sufficiently old, though not always remembered maxim, to extract by the mouth, whenever practicable, all balls lodged in the face.

The curious manner in which balls may be concealed in the bones of the face, and be discharged of their own accord, was shown in one instance in the Second Division after the battle of the Alma. A round ball had entered close to, but below the inner canthus of the eye, and being lost was not further thought of. The wound healed, and the patient had almost forgotten the circumstance, when, after suffering slightly from a feeling of dryness in one nostril, the ball fell from his nose, to his great alarm and astonishment, several months afterwards. It is somewhat singular that so little trouble should have been occasioned in this case, as it not uncommonly happens that a most distressing fetid suppuration attends the injury of bone in the region where this ball was probably lodged.

It is in wounds of the neck that the extraordinary manner in which the great vessels escape a ball's passage becomes most evident. Thus the neck has been injured by gun-shot, more or less severely, 128 times, and yet only 4 deaths have resulted from these wounds. Yet it must be true that a large number die on the field from those injuries. It would be useless, but sufficiently easy, to record cases in which balls, and even bayonets, have traversed the neck, and yet did not injure the great vessels; sometimes passing from side to side, sometimes from before backwards, it would appear almost impossible that the blood-vessels could have escaped the wounding agent, and yet no indication of any mischief followed. The great nerves suffer not uncommonly in gun-shot wounds of the neck, when such wounds are situated low down. Paralysis of the arm, setting in, either immediately after the infliction of the injury, or a few days later, affords evidence of such a lesion.

The soft coverings of the chest were wounded after April 1st, 1855, by gun-shot, more or less severely, 255 times, with 3 deaths resulting. In 24 cases, the bony, cartilaginous, or intercostal tissues were wounded, and one of these died. Lesion of the contents took place 16 times, although the ball did not penetrate, and 9 deaths resulted from that cause. The ball penetrated and lodged, or appeared to lodge 33 times and of these patients 31 perished, while in 9 cases the contents of the thorax were wounded superficially, 3 times with a fatal result. In 83 instances the contents were deeply perforated, and death followed in 71 cases.* It would thus appear that, with all our boasted improvements in the method of investigating the effects

---

* M. Legouest mentions, in a communication he has been good enough to send me, 6 cases of penetrating wounds of the chest, as having occurred in his divisions of the Dolma Batchi hospital at Constantinople, and of these the half died. Alcock gives 1 to 1 7-29, as the mortality attending his cases of penetrating and perforating gun-shot wounds of the thorax. In Guthrie's 106 cases, of whom a half perished, "the cavities were not penetrated. In the documents of the medical department I have found a record of 39 cases in which the chest was penetrated, and in some perforated by balls. In most of these there were signs of injury to the contents. Of these 39 cases, 27 died and 12 recovered. Meniere reports 20 cases of perforating wounds by gun-shot, many of them effected at very short range. All died, many very soon after being wounded. Nine penetrating wounds which he also mentions recovered, in all which there were signs of lesion of the lungs.

and progress of injuries of the lungs, the mortality has not abated much from what it has always been, when large numbers of men have sustained such injury from gun-shot. Wounds of the thorax are very common in battle when the combatants are in close proximity. This was particularly the case in the civil disturbances in Paris: and in siege operations the same holds good. The large surface, and elevated position of the thorax, accounts in some measure for this.

The distinction usually made between wounds of the parietes, and those which penetrate and injure the viscera of the cavity, is evidently a good one, as it separates between two classes of injuries of very different import.

Simple contusion of the walls may be caused by a spent ball, or by a ball which has impinged against some part of the soldier's accoutrements, and has thus been prevented from entering. Such an injury although not accompanied by any fracture, may yet be sufficient to give rise to hæmoptysis, severe constitutional shock, and internal inflammation. If the ball strike the edge of any of the metal plates which form part of the soldier's accoutrements, then the injury to the contents may be inflicted by the part so struck, as was the case in the following instance, in which a round shot was the missile, and the severity of the injury was little evidenced by the symptoms before death. Darling, private, 61st regiment,* was hit at Sadoolapore by a round shot, on the edge of the breast-plate, which was so turned inwards as to fracture the cartilages of the fifth, sixth, and seventh ribs on the left side, close to the sternum. The skin was not wounded. He walked to the rear, and complained but little for two hours, when he was seized with an acute pain in the region of the heart. His pulse became much accelerated, and he grew faint and collapsed. A distinct and sharp bellows'-sound accompanied the heart's action. He died in seventy-two hours from the receipt of the injury—the pain and dyspnœa, which had been so urgent at first, having abated for some hours before death. The heart was found to have been ruptured to an extent sufficient to allow of the finger being thrust into the left ventricle. The obliquity of the opening had prevented the blood escaping into the pericardium, which contained about two ounces of dark-coloured serum.

Dupuytren has drawn attention to the long period which ball wounds of the soft parietes of the chest take to heal, especially when they are "en gouttière." This he accounts for by the constant motion imparted to the walls by the movements of respiration.

If the blow from a ball be forcible, or strike directly on the chest without the intervention of any strong substance, then fracture of one or more of the ribs will probably be caused, and pos-

---

* Unpublished records of the Medical department.

sibly pleural or visceral inflammation, as well, from the effects of
the blow, or the presence of spiculæ driven inwards. These
fragments are at times long and sharp, and may be totally detached from the rib, and carried deeply into the lung substance.
The cartilage of a rib, although torn by a ball, is seldom driven
into the parenchymatous tissue, but remains so attached that its
fragments can be easily restored to their proper position.

It occasionally happens that a ball is arrested between two
ribs. This happened in the following case. Cassay, a private
in the 38th regiment, was admitted under my charge, into the
general hospital, on the 18th of June, suffering from a gun-shot
wound of the left side of the thorax. The ball, a large conical
one with a broad base, was much spent when it struck him. It
did not force itself into the cavity, but lay wedged between the
cartilages of the second and third ribs, on the left side, about an
inch from the sternum. On withdrawing the ball, the cavity of
the chest was found to be fairly opened, and the lung was visible
as it expanded and contracted. The patient had a severe attack
of pleurisy a few days afterwards, for which he was repeatedly
bled. Effusion, to a limited extent, followed, and his gums were
touched with mercury. For five weeks the wound continued to
suppurate freely. The lung became adherent to the parietes.
This patient had subsequently a short attack of bronchitis, but
ultimately made a good recovery. He went to England in
August, at which time he still complained of a severe pain in the
left clavicle and shoulder, which extended down to his hand, and
was attended by numbness and want of power. The pain was increased by touching the arm, and had continued since he was
wounded. In this case the cavity was opened, but the lung escaped injury. The non-collapse of the lung was well seen in
this, as in some other instances which fell under my notice.
The natural mode of repair, by adhesion between the lung and
the walls of the chest, and the troublesome affection arising from
injury to the nerves of the arm, were both illustrated in the
above case.

Pieces of shell, not unfrequently, open the cavity, but spare the
lung, while sometimes the reverse happens, and the lung may be
injured without the pleural sac being opened. The following
was a curious instance of this latter accident, without the thorax
being opened. The case occurred under the charge of my friend
Mr. J. H. Hulke, assistant-surgeon to King's College hospital, to
whom I am indebted for the details. Private Jeremiah O'Brien
was admitted into the general hospital on the 15th November,
1855, having been wounded by a piece of shell when the right
siege train exploded. His left arm and fore-arm were extensively shattered, and he had two small irregular wounds on the left
side of his chest, one just below the lower angle of the shoulder
blade, and the other on the same level, but about two inches
nearer the sternum. His breathing was quick and laboured, and

bright florid blood was bubbling from his mouth. His face was pale, his pulse flickering, and very feeble. He spoke with a firm voice, and begged his arm to be cut off. No communication could be detected between the wounds on the chest and the cavity within, but two ribs were found to be broken. His wounds were dressed simply, and his chest fixed. Beyond dressing, nothing was done to the arm, as he was not in a condition to undergo any operation. By night the breathing was easier, and he brought up less blood. Next morning his pulse was fuller, but intermittent. His spit still contained blood. His chest was naturally resonant as low as the fourth rib, but below this, by percussion and auscultation, dullness and friction sounds were discovered. He was cheerful, but, as he had not slept, half a grain of morphia was administered. He subsequently rallied somewhat, but died suddenly next afternoon, without any return of the bleeding. On examination after death, the sixth and seventh ribs were found fractured without displacement. The pleura costalis was entire. The part of the lung below the level of the fracture was entirely adherent to the ribs and diaphragm, while, in the upper part of the pleural sac, a small quantity of bloody serum was found. Opposite the position of the fractured ribs, the lung substance was extensively lacerated. A large rent ran inwards from the external surface towards the root, downwards towards the base, and upwards towards the apex. A large branch of the pulmonary artery was seen with an open torn mouth in the rent, while many other vessels stretched across it. The right or uninjured lung was ecchymosed at numerous spots on its surface, and in part emphysematous. Ecchymosed points were seen also on the surface of the heart and pericardium. The mitral valves, and endocardium of the left ventricle, were of a rosy hue. The segments of the tricuspid valve were bound together by a fibrinous clot, which narrowed the passage to the size of a small quill. Blood was found in the small intestines, but not in the stomach. Mr. Hulke remarks the arrestment of the bleeding by the mode in which the chief vessel was torn, as well as the conservative act of shutting off the rent in the lung, and the torn bronchi from the pleural sac by the formation of adhesions.

It is seldom that a conical ball will be found to lodge in a rib, as a round one has been seen to do, or yet to run round under the integuments, or at all to lodge within the chest. In fact, it very rarely fails to penetrate deeply, or pass quite through the entire cavity.

Non-penetrating wounds are more dangerous at some points of the thorax than at others. Thus, when a ball strikes a large bone like the scapula or the spine, or in those places where the large blood-vessels and nerves are situated, as in the axilla and upper part of the chest, the danger is greatly increased.

The gravity of penetrating wounds depends very much on their direction and their point of entrance, as when, with an incidence very oblique to the surface, they enter at some parts of the chest, they may traverse a portion of the cavity without touching the contents. So it happened in the following case. Fontaine, a private in the 90th, wounded on the 8th September, was admitted into the general hospital on the same day. The ball, after passing through the flesh of his left arm, which was at the moment in advance of his body, had entered the thorax in the axilla, and escaped at the inferior angle of the scapula, fracturing it, along with two of the ribs, at the place of exit. No immediate disturbance followed, but in twenty-four hours signs of acute pleurisy appeared, and required decided treatment. The ball had entered the cavity of the chest, but the substance of the lung had evidently escaped. Bone exfoliated by the wound of exit, which continued to suppurate long after that of entrance had closed. No bad symptom arose after the attack of pleurisy above referred to was subdued. I have seen this man lately in perfect health.

The finger is the only probe permissible in examining wounds of the thorax. If we thereby discover the projection inwards of fragments of a rib, or portions of it impacted in the lung, we should take immediate steps for their removal, even though the wound has to enlarged in order to allow of its accomplishment. The ribs are best fixed, and the wound left free, by means of strips of adhesive plaster passed from the spine to the sternum, and from above downwards, so placed as to embrace the wounded side only. Men wounded in the lungs require all the breathing space we can give them, and this is best managed by having the sound side free.

It is a singular circumstance connected with wounds of the walls of the thorax, that an intercostal artery is seldom opened. I neither saw nor heard of such a case during the war, so that we were spared the adoption of any of those operative procedures for its closure, which, Boyer remarks, are more numerous than the authentic cases of the occurrence of the accident.

Balls passing in front of the chest from side to side may cause very grave injury to the parietes, without absolutely wounding either the heart or lungs. This occurred in the following most interesting case :—

Fleming, a private in the 18th regiment, was admitted on the 18th of June into the general hospital, under Mr. Rooke. This lad was struck by a Minié ball, a little above the right nipple, as he stood sideways towards the enemy. The ball escaped below the left breast. The sternum was fractured and comminuted by the ball in its transit. Severe dyspnœa followed together with a slight attack of hæmoptysis. Repeated attacks of inflammation occurred over parts of both lungs, and the subsequent supervention of pericarditis necessitated bleeding and the use of tartar emetic, and subsequently of mercury, so as to touch the gums. The soft parts between the wounds of entrance and exit sloughed, and the sternum to the extent of about one and a half inches, together with the cartilaginous ends of the ribs thereto attached, came away in fragments, or were absorbed; so that by the 12th of July, a profusely suppurating wound had formed, 6 inches long by 2½ broad, across the front of the chest, laying open the anterior mediastinum, together with the right thoracic cavity, the opening into which was, however, sealed by the adhesion of the lung to the parietes. At the left extremity of the wound, and at its lower part, the heart was plainly felt only covered by the pericardium. A to-and-fro sound accompanied the motions of the heart, but these were not sufficiently pronounced to prevent the recognition of the two natural notes. Hectic fever, harrassing cough, and emaciation supervened. By the middle of July the wound had begun to granulate, and the patient seemed to improve. An attack of diarrhœa, however, prostrated his little remaining strength, and ultimately proved fatal. Before death, the pus with which the wound was filled receded on inspiration, and welled up when the lungs were emptied, as if it sank between the lungs when they expanded. On the morning of the day on which he died, a new sound was heard to proceed from the region of the heart, to which we never before heard any similar. It was exactly like the "clanking" note which accompanies the working of a pump when its gear is loose. There was the sucking in, and expulsion sound, together with this sharp peculiar note, which it is impossible to describe, but which immediately suggested the probability that the pericardium had been opened, and that the pus which filled the wound was alternately being sucked into and ejected from its cavity. On examination this view was confirmed, as a small hole was found at the inferior and left lateral aspect of the wound through which the pus appeared to be drawn in, and thrown out, during the action of the heart. After death, it was found that this aperture led into the pericardium, which was much thickened, and adherent to the heart, for a space of two inches by one, at the anterior and middle part of that organ. The opening mentioned led into a pouch formed by the pericardium round the roots of the great vessels, and which pouch communicated freely on the right side of the heart with the sac of the pericardium, at the base of the heart below the adhesion. Pus was freely effused into the pericardium, and the surface of that membrane, as well as that of the heart, was of a drab colour, and thickly coated with lymph of a low type of organization. The heart itself was healthy. The lungs were somewhat congested, and their anterior surfaces were adherent to the parietes. The coats of the stomach were unhealthy, but beyond this nothing was observed.

The noble struggle made against death by this poor boy, the very extensive injury, the opening of the pericardium, and the sealing of both sides of the thorax by the pleural adhesions, were all points of much interest and no little instruction.*

The two following cases show how small a difference in the place of transit of the ball may determine the question of life or death:—A Zouave was struck at the Alma by a round ball, which entered the parietes close to the right nipple, and escaped at a corresponding point on the left side. The ball passed in front of the sternum, which it fractured. Curiously enough, no inflammation whatever of the contents of the thorax followed, and he was in a short time discharged well. The points of entrance and exit differed little in this and in the case of Fleming; but the projection of the sternum being less in this patient, the result was very different.

A Russian soldier lay close to the Zouave just referred to, who, in the same battle, had been struck by a ball about a quarter of an inch to the outside of the right nipple. The ball had then passed behind the sternum, fracturing it badly in its course, and escaped close to the left nipple. Double pneumonia and pericarditis followed, and he died. The whole contents of the thorax were found implicated in one vast inflammation, Not being present at the post mortem examination, I did not learn how far the pleuræ or pericardium were injured (as I understood they were) primarily.

When a ball fairly enters the chest, and either penetrates or traverses the lung, the danger is most imminent. These injuries, however, are not so fatal, on the whole, as similar wounds of the head or the abdomen. The younger Larrey and Menière both record the circumstance, that the majority of the killed in the civil commotions of 1830 in Paris, succumbed from penetrating wounds of the thorax. The immediate danger will depend upon the depth of penetration, and the part implicated. If the heart or great vessels are wounded, death will in general be instantaneous. When the lung is only superficially wounded, then the vessels which are injured must be of small calibre; but the deeper the ball penetrates, the larger are those encountered, and, consequently, the more mortal is the wound. The patient may be suffocated at once by the blood, or it may escape in such quanti-

---

* John Bell (2nd Discourse on Wounds, p. 302) refers to a case related by Galen, in which part of the sternum was removed, the pericardium opened, and the man cured. He thus comments upon it—" Here, then, we have, upon that authority which has been always respected, a case exceeding in the miraculous all that has ever been recorded by the patient Vander Wiel, or gathered by Schenkius, or any German commentator among them,—a man with a slow suppuration, confined matter, a carious sternum, and the heart absolutely exposed and bare." In Fleming's case we had all the unfavourable symptoms, but unfortunately not the recovery.

ty as to cause death, within a short time, by exhaustion. If the wound be at all severe, the shock is very great, and blood generally passes from both the mouth and the wound. That from the mouth is frothy, while that from the wound is darker-coloured in general. The wound being high in the walls of the thorax, will make the escape of blood by the orifice less in quantity than if it be situated low down, and such situation will render the evacuation of the effused blood, or serum, more difficult afterwards. Air, as well as blood, will generally escape by the wound, and thus the presence of these two signs—blood by the mouth, and blood and air by the wound—are unequivocal proofs that the lungs have been injured, although their absence does not prove the opposite.

The dangers which attend a penetrating wound of the lung, are thus, primarily, hæmorrhage and collapse, as well as those from suffocation, if the bleeding be profuse. The hæmorrhage and the fainting are, by a sort of paradox, both the patient's danger and his safety. Secondarily, the danger of such wounds proceeds from inflammation and its products, the exhaustion which attends prolonged exfoliations and suppuration, together with that which arises from the organic diseases that are thereby so apt to be engendered.

A short, tickling, harrassing cough, attended by bloody expectoration; a cold and bedewed surface; a pale anxious face; a weak trembling pulse; palpitations of the heart; oppressed breathing, arising in the first instance, according to Hunter, from the pain occasioned by the action of the wounded lung and muscles, and afterwards from the inflammation and effusion—these are the usual symptoms which attend penetrating wounds of the lungs. At a later date, if the bleeding cease—a circumstance which will be evidenced by the disappearance of the collapse, the return of the heat to the surface, and of strength to the pulse, as well as by the length of time which has elapsed since the infliction of the wound—then those symptoms which result from inflammation appear. We have thus two stages or periods which demand separate attention in our treatment—that during which there is internal hæmorrhage with collapse, and that which follows and is accompanied by reaction and inflammatory action; to these I might also add that of convalescence.

The collapse which follows penetrating wounds of the lung, though dangerous, is yet, if not very profound or prolonged, the best guarantee for the patient's safety. To such cases the observation of Hewson is peculiarly applicable: " Languor and faintness being favourable to the congelation of the blood, and to the contraction of the bleeding orifices, should not be counteracted by stimulating medicines, but, on the contrary, should be encouraged." With our modern notions on bleeding, it is often difficult to reconcile the necessity, which experience shows there is, for energetic depletion when reaction sets in. The majority of our patients were certainly not subjects in which this remedy could be pushed so far as Guthrie and Hennen would appear to recommend; but I think it was very generally observed that those cases did best in which early, active, and repeated bleedings were had recourse to. It is well known, that in sieges generally, soldiers do not show their usual tolerance of bleeding, and when their health is so much undermined as it was at Sebastopol, the surgeon is often placed in a most unpleasant dilemma. That many most excellent recoveries were made without having recourse to the lancet is undoubtedly true; but not a few, I fear, died from want of it. When the loss of blood by expectoration and by the wound has been very free, of course the necessity for abstracting it otherwise will be much less. The system is then far more easily reduced to that which favours the formation of the "caillot tutelaire." We must, in cases where venesection is required, be especially careful to bleed by a large orifice, and be guided by effects.* This, with perfect rest, the lowest diet, cooling drinks, and possibly digitalis, must form our means of managing the early stage. Any return of the oppression will show the necessity for further depletion. In wounds from gunshot, the patient should be allowed to lie in the position which he chooses; but if the wound be a stab, the position prescribed should be that which will favour the adhesion of the pleuræ; and when there is effusion within the thorax, that which will allow of its escape.

To determine whether the blood which flows from a wound in the thorax proceeds from a wounded intercostal or from the lung, has called forth more acumen and research than it would appear to merit. The difficulty will be greatest when a knife has been the instrument, and the wound made is very oblique. In large wounds, Sanson lays down the following means of diagnosis: 1. Whether the blood be arterial or venous; 2. By turning out with forceps the lips of the wound, and seeing whether the blood proceeds from one of these lips; 3. By compressing the superior lip of the wound with the finger, *i. e.* pressing upon the inferior border of the upper rib, where the wounded intercostal may be placed. He objects to the use of a roll of card introduced in the shape of a gutter, because when that can be done we may be able

---

* "Until the danger of immediate death from hæmorrhage is over," says Hennen, "we must not think of employing anything except depletion by the lancet; it, and it only, can save the life of the wounded man." "It is only by these repeated bleedings," says John Bell, "that the patient can be saved. The vascular system must be kept low in action, and so drained as to prevent the lungs from being oppressed with blood. One thing is very clear," he adds, "that if the surgeon bleed only when the cough and bleeding from the lungs return, he never can do wrong.'

to see the wounded vessel with the eye; but the examination of the wounding instrument will often show whether it could penetrate deep enough to injure the lung.

Bleeding from the lung makes itself apparent by both rational and physical signs. Some of these are common to all hæmorrhages, external or internal, while others are present in intrathoracic effusions of whatever description. Of the rational signs, paleness of the face, coldness of the surface, a small, concentrated, and quick pulse, giddiness, and syncope are those referable to the loss of blood; while the dyspnœa,* sometimes amounting almost to suffocation, the feeling of weight in the chest, the anxiety, restlessness, and the decubitus on the wounded side belong to all effusions. The physical signs are also common to all effusions. They are—a dilated chest, little moved during respiration, bulged intercostal spaces, dullness on percussion, and the absence of vesicular breathing. If there be air also present, we will have added those signs which are peculiar to such a complication, and which are recognizable by percussion and auscultation. The peculiar ecchymosis described by Valentin, and which results from the escape of blood into the subcutaneous cellular tissue, seldom appears; but if it does, it is according to many a valuable sign of hæmorrhagic effusion.† If, then, after a gunshot wound of the thorax, we have those signs present which would indicate the loss of blood, as well as those which indicate the existence of fluid in the pleura, embarrassing the functions of the contained viscera, the diagnosis is plain. If blood escape by the external wound during respiration, or after a cough, the opinion will be strengthened that blood has been poured out, and occupies the pleural sac.

The danger from hæmorrhage is greatest during the first twelve hours, and is pretty well over by the second day. A flow may however continue, in greater or less quantity, for eight or ten days, but then it is seldom to any serious amount. If the

---

* Sabatier mentions having seen patients perish of hæmorrhagic effusion in whom the breathing was not disturbed, and who could lie in any position.

† Lucz remarks upon this point—" Valentin pretends that the ecchymosis which is observed *on the loins, in wounds* of the thorax, is a pathognomonic symptom of effusion into the pleura, and that its absence is a counter-indication to paracentesis. Larrey says he constantly observed this fact, as do many other practitioners, such as Louis, David, &c. However, after the observations collected by Degranges, Chaussier, Callisen, Saucerotte, and others, we cannot look upon this phenomenon as a certain sign of hæmo-thorax; because, in many circumstances where the effusion really exists, it has not been observed, and it has followed non-penetrating wounds.

quantity of blood effused be small, it will probably be absorbed; but if it is in large quantity, and especially if air is also present, the gravity of the lesion is much augmented. So soon as all fear of a renewal of the bleeding is over, the effused blood, if in quantity, should be evacuated by operation; but, as Sanson says, it is better to be a little late than too early in taking this step.

There is no question connected with wounds of the chest so difficult to solve, as that which has reference to the management of internal hæmorrhage. The embarrassed state of the lung demands the evacuation of the fluid, and yet, if we allow it to escape, the bleeding from the lung is renewed, and death results. So it was in the following case:—

Hannihan, a private in the Royal Irish regiment, was admitted into my wards in the general hospital on the 18th of June. While lying on the ground, with his head towards the enemy, he was struck above the left clavicle by a rifle ball, which traversed his lung from its summit to its base, and was found lying quite superficially in the left lumbar region, from which position it was removed. The dyspnœa, on admission, was very great, and the hæmoptysis most profuse. The surface was cold, and bedewed with cold perspiration. The pulse was weak and tremulous, and the decubitus was on the wounded side. The removal of the ball was followed by a tremendous gush of blood from the incision made, and the blood continued to flow in such quantity that I had to close the wound to prevent immediate dissolution. The necessity of guarding against a suddenly fatal event, was for the moment paramount to the indication of freeing the embarrassed lung of the effused blood; and as the hæmorrhage, moreover, appeared to be active, I wished to try to check it by the pressure which would result from the blood being allowed to accumulate in the thoracic cavity. The patient was twice largely bled, and he had acetate of lead and opium given him. These measures appeared to afford him some relief. Next day he had rallied considerably. His pulse was better, and his look was less distressed. By the afternoon of that day, the dyspnœa became so urgent that I allowed a considerable quantity of the collected blood to escape. This gave him, for a time, decided relief. The severe exhaustion which, however, soon followed this step, and the return of the dullness on percussion to its former level, seemed to intimate a renewal of the hæmorrhage; hence I did not reopen the wound, but determined to abstain from all interference till the bleeding vessel had had time to close. The patient was so completely prostrated by the hæmorrhage which had evidently taken place internally, that I could not have recourse to any further depletive measures. The stethoscopic examination of the chest discovered amphoric breathing over the upper part of the left lung, while over the whole surface of the right chest the respiration was harsh and loud. Dullness existed on the left side from the base of the lung up to an inch and a quarter above the level of the nipple. There was suppression of urine for thirty hours after admission. This patient died on the fifth day, without any change in his symptoms from those noted above. The left side of the thorax was found more than half full of blood, for the most part fluid. The lung was half solidified and compressed against the spine. Lymph was effused to a limited extent on its surface. The ball had traversed the lung in a direction from above downwards and backwards. Its track was ragged and coated with lymph. The three upper and the three lower ribs were fractured. The patient's back, on the wounded side, was ecchymosed before death, and gave him much pain. This discoloration bore much resemblance to that ecchymosis described by Valentin; only it appeared at too early a period, and was not sufficiently pronounced to accord with his description.

I am not in a position to determine whether the retention of the blood in the cavity can really exert so great a pressure on the wound in the lung as to arrest the bleeding; but such was the opinion of Valentin, Larry, Sanson, and Dupuytrer. I am disposed to think that, in such cases as the foregoing, it would be better practice to open the cavity freely by enlarging the wound, so as to allow the blood to escape freely, and thus favour the contraction of the lung and the closure of the vessel; but in Hannihan's case such a step would have been attended with much danger, from his great prostration.

If the lancet be employed in such cases, it is a matter of the greatest nicety, and requires the utmost discrimination and judgment, to abstract exactly the quantity of blood requisite for producing the desired effect without exhausting the patient, whose system has been already so much drained by the internal hæmorrhage.

Hæmoptysis does not always occur in penetrating wounds of the lungs, and dyspnœa may be but slightly marked at first. The following case was an example of this:—M'Kennah, private 77th regiment, was admitted into the general hospital, July 27th. When in one of the advanced trenches, a Minié ball struck him obliquely from the left side at the middle of the supra-spinous fossa of the left scapula, and lodged. On admission, a couple of hours after the receipt of the wound, slight dyspnœa was the only observable symptom, and the only thing the patient himself complained of. The finger passed into the wound showed the direction of the ball to have been towards the centre of the body, but nothing was detected except some roughness along the posterior border of the scapula. In the evening the dyspnœa was more marked, and the pulse had increased in frequency. The decubitus was dorsal throughout. Emphysema appeared over the surface of the right side of the chest. He was largely bled. Next day the above symptoms were notably exaggerated, and dullness was added on percussion on the right side, posteriorly and laterally. The respiration was puerile over the anterior superior half of the right, and over the whole of the left lung. The bleeding was repeated, digitalis ordered, and nothing allowed in the way of food but milk and cold tea. On the 29th, the dullness had invaded the inferior and lateral aspect of the left lung. The dyspnœa became very urgent, and was not relieved by any treatment, depletory or otherwise, and he died on the 30th. Fluid blood, seemingly the product of oozing, was found in both pleural cavities, and some air also existed on the right side. Both lungs were much diminished in volume, and floated towards the upper part of the cavities. The ball had passed through the second rib, near the posterior superior angle of the scapula, and perforated the apex of the left lung with a transit of one and a half inches. It had there pierced the body of the second dorsal vertebra, fracturing and partially displacing forwards its anterior half. It had then entered the right pleural cavity, traversed the apex of the right lung, struck and fractured the second rib on the right side about its centre, and finally fell spent within the pleural cavity. The lungs were gorged with blood, and their outer and inferior surface were coated with lymph. If one lung only had been wounded, the ball and the effusion might have been both got rid of by operation; but when both lungs were implicated, such interferences would only have hastened death.

The emphysema which was present in this case, was probably due to the oblique direction of the wound. It was a very rare occurrence in the chest wounds which I had an opportunity of witnessing.

The inflammation which follows gun-shot wounds of the lungs, requires the same treatment as that which is given to inflammation from any other cause. When only a small part of the lung has been penetrated, then the pneumonia may be at first localised; but it will soon spread if not promptly subdued. During convalescence, the great point which demands attention is to guard against all sources of relapse, as inflammation is very apt to be re-established, and if it does reappear, the danger of its giving rise to purulent effusion is very considerable. Serous effusions often cause much annoyance in cases of wounds of the chest. According to Guthrie, such effusions take place, in general, from the third to the ninth day, and, if large, imperatively demand early evacuation. I fear this rule was not always attended to during the late war. It is difficult to know what is the best period of the disease to put it in practice.

The strictest regimen should be maintained for ten days or a fortnight after the infliction of a gun-shot wound of the lung. Any irregularity in diet, or indulgence in ardent spirits during convalescence, it is most apt to cause dangerous, if not fatal relapses. Not a few were lost in the East from such carelessness. Opium is of much use in allaying the troublesome cough, which often continues for a long time. Hennen speaks of "a sense of stricture and considerable pain in raising the body to an erect posture, with great anxiety on walking up an ascent," as being frequent consequences of gun-shot wounds of the chest; and at another place he says, "diseases which, although we cannot call them pulmonary consumption, agree with it in many points, particularly in cough, emaciation, debility, and hectic, are often the consequences." Veritable phthisis has, however, as is well known, been cured by the rough medication of a gunshot wound. We had no opportunity of watching the remote results of these wounds, as the patients passed from under our care too soon for their development.

Of wounds perforating both sides of the chest, I met with four examples only. In all these the wound was inflicted by grape, and all died in a very short time.

Balls are well known, occasionally, to become sacculated in the lung. This circumstance, as well as the very small amount of irritation which the presence of such a body may give rise to, was illustrated in the following case. The case was first related to me by my friend Deputy-Inspector-General Gordon, C.B., and I afterwards found the particulars of the early symptoms in the medical reports of the regiments serving in India:—A soldier of the 53rd, serving in the Punjab, received a ball on the left side of the thyroid cartilage, which coursed round the neck, entered the apex of the right lung, traversed it to near its base, and lodged. Violent dyspnœa, urgent cough, and bloody sputa followed. The patient, from the fear of suffocation, could not lie down for several days. These symptoms were allayed by treatment, and in two months the man was discharged, feeling no inconvenience from his wound. This patient died six months afterwards of a contagious fever, when the ball was found closely sacculated in the lower lobe of the lung, at the apex of which a small puckering was seen, but no trace could be discovered of the ball's track from the apex to its place of sacculation. The lung was free of disease. In the following case the position of the ball was not discovered:—A soldier of the Buffs, wounded on the 8th September, received a ball on a level with, but slightly external to, his right nipple. Profuse hæmoptysis, fainting, great dyspnœa, oozing of blood from the wound, and the escape of air followed. He was largely bled, and his symptoms thereby relieved. Ten hours afterwards, a return of the difficulty of breathing called for further depletion, and the use of antimony. Pneumonia followed, which implicated the lower half of the wounded lung. The treatment was that for pneumonia generally. The wound suppurated, and ultimately closed. When the patient left the hospital in December, the lung acted well throughout, except for a short distance round the wound, where it was dull on percussion, and seemingly impervious to air. The vocal resonance was notably increased over the upper part of the wounded side of the thorax.

The direction taken by the ball, and its position as found after death, give interest to the following case:—At the Alma a soldier was struck by a musket ball, on the outer side of the left shoulder. His arm was by his side at the moment he was wounded. It was observed that the ball had passed through the head of the humerus, but its ultimate position could not be ascertained. Nothing was done for the arm. The ball was supposed to have made a clean hole through the bone. A severe attack of pleurisy followed, and on the subsidence of this, pus was found to point both below the clavicle, and in the axilla of the wounded side. Much bone came away. Pus flowed copiously by the openings which were made in the axilla, and below the clavicle. The patient became hectic and died. It was then found that the ball, having passed through the head of the humerus and the glenoid cavity, had entered the chest between two of the ribs, and having run forwards within the cavity, and between the walls and the pleura, had lodged in the anterior mediastinum, where it was found coated with lymph. The chest symptoms, the surgeon in charge informed me, had been very slight, and the presence of the ball had given rise to no uneasiness. If the joint, which was the main source of irritation and hectic, had been excised early, a more favourable result might have followed.

The four following cases are further illustrations of most severe gun-shot wounds implicating the lung:—

At the Alma a soldier was struck by a ball near the centre of the left axilla. The bullet escaped on the same level as that at which it had entered, and within an inch and a half of the spine. Profuse hæmorrhage by the wound and by the mouth followed immediately, and caused the patient to faint. He was bled at night, as well as next morning, to relieve the dyspnœa, which was urgent. A severe attack of pneumonia followed, which, though subdued, recurred on two subsequent occasions. By December the lung had recovered, except at its base, where it was impervious to air. The respiration at the summit was exaggerated. There was in the hospital, at the same time, another man, whose wound and its results were exactly similar, only that the ball had entered by the right axilla in place of the left, and had escaped a very little lower than in the last case. In this case the liver escaped injury.

A sergeant was struck at the Alma by a musket ball, on the right side, between the sixth and seventh ribs, close to their angles. The ball traversed the lung, and escaped close above the inner angle of the clavicle of the same side. The man said that, on the receipt of the wound, his mouth filled with blood, and that he fell down and thought he was killed. Profuse hæmoptysis continued for some days after his admission into hospital. He was largely bled a few hours after being wounded, and also on the two succeeding days, when the difficulty of breathing, from which he suffered, became severe.

Tartar emetic was given him, and he was kept exceedingly low for several days. Both wounds suppurated freely. Amphoric breathing was very evident over the upper part of the wounded lung; but there was no marked change on percussion anywhere, for a week after the receipt of the injury. He complained of severe pain in the injured lung during the whole period he continued in hospital. Three weeks after being wounded, there was a deficiency in the respiratory murmur all over the right side, which deficiency was balanced by an increase on the left. Bronchophony was marked at the upper part of the right side. There was dullness now on percussion all over the right lung, but chiefly at its upper part. The expectoration was profuse and purulent. Cough severe and painful. Pulse high and irritable. His gums were sore with mercury, and blisters had been repeatedly applied to the surface of his chest. He gradually recovered, under the influence of a generous diet; and when he went to England, about four months after being wounded, both wounds were closed, the anterior having cicatrized first. At that period the right side of his chest was somewhat contracted and flattened. The respiratory murmur was fair over the upper two-thirds of the right lung, but faint towards the base. Percussion gave a normal note, except at a small point just at the apex and at the base, where the sound was dull. A good deal of bone had been discharged by the wounds during convalescence.

A French soldier had a Minié ball driven through his right chest at Inkerman. It entered an inch below the nipple, between two of the ribs, and escaped behind, exactly opposite the place of its entrance, within two inches of the spine, fracturing one rib, and chipping another. Severe hæmoptysis and bleeding from the wound followed. He was bled frequently, and kept very low afterwards. Most violent inflammation set in, and effusion took place into the pleural cavity. The fluid was not evacuated; but while it was being absorbed, the wound of entrance having closed, a most violent and prolonged attack of trismus seized him, which, for a couple of days, threatened to cause death, but which ultimately yielded to large doses of opium, without the spasms becoming general over the body. This patient perfectly recovered, and was sent to France.

A soldier of the Guards was struck at Inkerman by a rifle ball, which was fired at a short distance behind him by one of our own men. It entered below the angle of the right scapula, and escaped between the fourth and fifth ribs, chipping the upper edge of the latter. The hæmoptysis was very profuse, and much blood escaped by the wounds. He sank down exhausted, almost immediately on receipt of the wound, and lost consciousness shortly afterwards. He lay a considerable time, he could not say how long, before he recovered. When he was received into hospital, blood continued to ooze from his wounds, he spat constantly, and his breathing was greatly impeded. He was bled twice during his stay in the Crimea, and when I saw him, a month afterwards, he had in a great measure recovered. The exit wound had closed, but that of entrance had taken on a phagedenic action for some days, and was not yet healed. The lung acted well; he could lie on either side; and, to all appearance, he was in a fair way to a complete recovery.

When no adhesions are formed, by which the ball or other foreign bodies driven into the thorax are arrested, they generally are found lying on the diaphram, in the angle formed by it and the costal walls, and close to the vertebral column.

The track of a ball through a lung has been occasionally found to become fistulous, becoming lined by a membrane, and containing curdy pus. The pulmonary tissues around these tracks becomes indurated, and they may, or may not, have an orifice to the exterior of the chest. A circumscribed abscess may exist between the ribs and the lung, or be in the lung substance itself, and communicate with this track. The perfect manner in which these collections, and the track connected with them, are closed off from the lung, and the evil which may arise from the presence of this pus, make it a question, which the facts before me do not enable me to discuss, whether, or not, it would be advisable to evacuate it by operation, seeing that our modern means of diagnosis would permit of its detection. This evacuation could be accomplished by such a puncture through the parietes, as would insure the closure of the wound as soon as the object was effected.

CHAPTER VI.

GUN-SHOT WOUNDS OF THE ABDOMEN AND BLADDER.

The returns of the war, after April 1st, 1855, show flesh contusions and wounds (simple and severe) of the abdomen, among the privates, as having occurred 101 times, with a fatal issue in 17 cases. There were 38 penetrating wounds with lesion of viscera, and 36 deaths in consequence; while 65 times the abdomen was perforated, and 60 deaths resulted.* Four cases of rupture of viscera without wound were fatal.

The abdominal cavity, from the want of a bony protection in front, as well as from its large surface, is very liable to severe injury in battle, and there is no cavity in the body the injuries of which are more serious, or more often fatal. The ribs protect the contents of the thorax from contusions, and wounds from pieces of shell often fail to injure either the lungs or heart; but when a projectile impinges with any force on the abdomen, the effects are seldom limited to its walls.

It is often difficult to tell what influence a certain wound will produce when it affects the abdomen. At times an accident apparently severe is followed by trivial consequences, while the most disastrous results may arise from an injury which shows little external indication of its severity.

Contusions by round shot are among the most dangerous injuries to which the abdomen is exposed. The hollow or the solid viscera, as is well known, may be thus ruptured, and rapid death follow, without much external sign of so severe an accident. Every campaign furnishes examples of this. A contusion may, however, arise from a less ponderous missile than a round shot,

---

* M. Legoust mentions 3 cases of penetrating wounds of the abdomen in the Dolma Batchi hospital, all of which died. Alcock reports 19, only 1 of which recovered. Meniére mentions 14, in which the ball penetrated, 2 of them being through the side, and all died; while of 7 others, in which the ball passed through the side only recovery followed. In the Indian wars I find the record of 38 penetrating or perforating wounds of the abdomen, of whom 32 died and 6 recovered. Colles states, that in the sieges of Moultan "not one case recovered in which the abdomen was fairly shot into and the small intestine wounded." Sedillot tells us, that in the expedition against Constantine, they lost all those whose abdomens were penetrated by gun-shot.

and the injury be not so serious. The state of tension of the wall of the abdomen at the time of the accident, appears to exercise no little influence on the effects produced. When a man is lying on the ground, and the muscles are completely relaxed, then the injury inflicted on the contained viscera may be very severe; but if the muscles are in action and tense, then the force of the blow will be somewhat mitigated. At least such is the only manner in which I could explain several anomalous cases that fell under my notice.

Vomiting and pain in the abdomen are the signs of injury to which contusions of the cavity generally give rise; and if no serious damage has been done, all the treatment those cases require is such as will ward off peritoneal inflammation, which may steal on very insidiously. If any internal rupture has taken place, we can do little to prevent a fatal issue.

Shell wounds of the walls of the abdomen are very commonly followed by extensive sloughing, and the danger of the morbid action laying bare the intestines, or at any rate favouring their subsequent protrusion, is considerable. In one case which fell under my observation nearly the whole of the anterior wall of the abdomen was destroyed by the sloughing caused by a shell wound.

Guthrie seems to think that a greater amount of destruction occurs in the abdominal walls, than can be accounted for by their mere injury; this loss being probably caused by their absorption.

Balls often traverse the abdominal walls for a considerable distance without entering the cavity, and they do this at times by so long a transit as to describe half the circuit of the body. Of this very many cases occurred in the Crimea. The strong aponeurosis which protects the front of the abdomen, exercises a great influence in deflecting the ball when it has struck at all obliquely. The track which is thus made requires careful management during cure to get it to close. If it be long, it is good practice to make a counter-opening at its centre, in order to prevent the lodgment of pieces of cloth or pus in its interior. This can, however, be necessary only when, neither by syringing nor by the introduction of an elastic bougie, we can get quit of them.

Abscesses among the muscles are not uncommon, although very disagreeable complications of gun-shot injuries, and especially of contusions of the abdominal walls. Severe pain, vomiting, and other symptoms which may be mistaken for those of internal inflammation, may be thus set up.

If the amount of inflammation caused by contusion or other injury of the abdominal wall be limited, then adhesion will take place between the parieties and the omentum or viscera, and will afford a great safeguard against the effusion of blood or other matters, into the cavity. If, however, the paries in part slough, so that the gut is laid bare or opened, the injury is one of great gravity.

It is sometimes very difficult to say whether a ball has perforated the abdomen or not. The relative position, and even the peculiar characters of the two orifices, will not guarantee a decided opinion. Far less can we say, from the apparent direction of the wound, that any of the viscera have been injured. It is neither allowable nor desirable that we should make such a search as will determine the question; for if the ball be not easily found, we never "amuse ourselves," as Le Dran expresses it, "by seeking for it," and the treatment ought to be such as will provide for all contingencies. In the following case, the ball appeared not only to have perforated both the abdomen and the chest, but also the diaphragm; yet probably it ran merely under the the integuments, possibly traversing the diaphragm close to its anterior border, and wounding none of the abdominal or thoracic viscera. A ball struck a French soldier just above the crest of the ileum, and about four inches from the spine. It escaped close below the inner end of the clavicle on the same side. At the time he was struck this man was on his knees, as he was in the act of rising from the ground on which he had been lying. He had hiccough and considerable prostration for three days, and also an attack of pleurisy, all of which he had recovered from a fortnight after injury, when I first saw him.

The fatality of penetrating wounds of the belly will depend much on the point of their infliction. Balls entering the liver, kidneys, or spleen, are well known to be usually mortal, although exceptional cases are not rare. Wounds of the great gut are also always recognized as much less formidable than those which implicate the small. Thomson saw only two cases of wounds of the small gut, after Waterloo, in the way of recovery; but Larrey reports several. Gun-shot wounds of the stomach are also exceedingly fatal. Baudens records a remarkable case of recovery, although complicated with severe head injuries. The syncope which followed the severe hæmorrhage in this case lasted for ten hours, and doubtless assisted, along with the empty state of the stomach at the moment of injury, in preventing a fatal issue.

The extraordinary manner in which, not only balls, but also swords and ramrods, may traverse the abdominal cavity, and yet not wound any viscus, has been often dwelt upon by military surgeons. The escape of the viscera in the following case, which occurred in India, was most remarkable. A soldier of the 28th regiment endeavoring to commit suicide, leant over his musket, and drew the trigger with his toe. The ball passed into the abdomen, on a level, but a little to the left, of the umbilicus, and escaped through the centre of the crest of the left ileum behind. He died in a month. The intestines were found matted together, and large portions of them were gangrenous, but no perforation of the gut could be discovered. The surgeon, Dr. Young, adds n his report,—" This examination, however, in some particulars unsatisfactory, has at least established the fact that the intestines were not perforated by the ball; but how they escaped defies any conjecture I can form on the subject." In another case which occurred at Meanee, the ball was ascertained to have gone fairly through the abdomen, yet not to have injured any of the viscera. It is impossible, however, to be certain of such a circumstance, unless an after-death examination verify a supposition we are too apt to form.

The just and perfect support afforded by the abdominal viscera to one another, and the manner in which they fill their containing cavity, supply a safeguard against effusion after wounds, which has ever been the astonishment and admiration of observers. The smaller and less torn the wound in the gut is, the more likely is this favourable result to occur. Littre's celebrated case of the madman, has ever served as the type of such wonderful acts of "conservative effort." The pressure, too, favours that adhesion between the viscera which is so potent a preservative against evil.

The following case, reported by Dr. Taylor when surgeon of the 80th, affords an example of a gun-shot wound injuring the smaller gut, while at the same time it shows the effects of such a wound, and also the state of the parts a considerable period after the infliction of the injury. It is taken from the Records of the Medical Department:—

Private Paul Massy was shot through the abdomen at Ferozeshah. Very slight symptoms followed, so that it was supposed the ball had coursed round the cavity, and had not penetrated. He mentioned having passed some blood in his stools after receiving his wound. The ball had escaped near the spine, having entered in front. He recovered slowly, but perfectly, except that he continued subject to bowel complaint, and finally died of spasmodic cholera, a considerable timo (exact period not specified) after being wounded. For a year before death he was almost constantly under treatment for dysentery. When examined after death the following was the condition found. I give it in Dr. Taylor's own words. "Cicatrix of a gun-shot wound in the left linea semilunaris, about four inches above the crest of ileum; and on the same plane posteriorly, another cicatrix an inch to the left of the spine. Omentum firmly adherent to the internal surface of anterior cicatrix, and gathered into a fold or knot at that part. The intestines were neither there nor elsewhere morbidly adherent; but the fold of intestine immediately opposite to the cicatrix presented a line of contraction, as if a ligature had been passed tightly round the gut. The fold of intestine immediately above presented the same appearance, and on the first fold, four inches from the first-noticed contraction, and in a line below the umbilicus, was another similar appearance. These three contracted places were of a darker hue and more vascular, than other portions of the small intestine; having, however, throughout an arborescent vascularity, and being in the sodden state constantly seen in sudden cases of spasmodic cholera. The mucous membrane of the small intestine was generally of a pale pink colour. No ulceration of the large gut. Upper part of the colon attenuated, and contracted *in situ*. Rectum thickened."

When a ball merely enters the gut, it may be thrown out by stool. Such a case occurred in the 19th regiment in the Crimea, and is reported by the surgeon in the *Lancet*, vol. 1, 1855.

If a vascular viscus be wounded, or a large blood vessel opened, then hæmorrhage may take place within the abdomen to a very serious and fatal extent. The mutual pressure of the viscera does much to prevent bleeding from the former source, and the lax attachment of the arteries in general enables them to escape. If blood be poured out suddenly and in quantity, it will partly escape by the wound, and partly collect at the most dependent part of the abdomen, or in the pelvis. Baudens mentions as a certain sign of a quantity of blood being collected in the pelvis, the incessant and insupportable desire to micturate caused by the pressure on the bladder, and which is set up although there is no urine in the viscus. Besides the immediate danger which proceeds from the loss of blood, such effusions, if in quantity, fail to become absorbed, decompose, set up inflammation, and cause death. The quantity must be small which will insure its absorption. It is therefore a matter of some importance to evacuate such accumulations by reopening the wound, rather than.to attempt its removal by operation afterwards

The symptoms of penetrating wounds of the abdomen are those which belong to the accident proper, and those which result from its consequences. The collapse is generally very severe, and this is the case, too, in many instances in which the injury appears at first very superficial and trivial. While, in general, this shock and alarm are indicative of deep and serious lesion, they are often excited by no apparently adequate cause. If some hæmorrhage, or the effusion of any of the secretions, as bile, or the contents of any of the hollow viscera follow the injury, then the collapse will not only be severe, but will continue.

The subsequent symptoms of these wounds will partake of two characters—those common to all inflammations of the abdomen, and those arising from the inflammation of the particular organ injured. The inflammation which is so certain to occur in the peritoneum requires very careful watching, as it often sets in very slowly, and deceptively. "The consciousness of imperfection induced in the cavity," of which Hunter speaks, makes it peculiarly apt to take on an inflammatory action.

The position and direction of the wound, and the concurrent symptoms referable to the lesion of special organs, will lead us to surmise the injury of this or that viscus. The persistent vomiting, the ejection of blood by the mouth or by stool, or with the urine, the escape of special secretions, as bile by the wound, the peculiar pain or sensation experienced by the patient, will be our chief indications in determining the part hurt.*

The treatment of simple, non-penetrating wounds requires but little notice,—the prevention or subdual of inflammation, and the favouring by position of that conservative adhesion between the viscera and the parietes which is desirable if sloughing should set in, so as to endanger the opening of the cavity.

The management of penetrating wounds is not much more difficult, but the results are very much less satisfactory. When the penetration has been occasioned by a ball, it is not often that we have an opportunity of verifying the fact of viseral lesion. No attempt should be made to follow the ball. The wound should be lightly covered, the patient placed in such a position as will relax the abdominal walls, fomentations applied by means of the lightest possible material, opium freely given by the mouth; and, if inflammation set in, then leeches and even general blood-letting may be had recourse to.

"All wounds that enter the belly," says Hunter, "which have injured some viscus, are to be treated according to the nature of the wounded part, with its complications, which will be many; because the belly contains more parts of very dissimilar uses than any other cavity of the body, each of which will produce symptoms peculiar to itself and the nature of the wound." "It cannot be too frequently repeated," says Dr. John Thomson, "that copious blood-letting, and the use of the anti-phlogistic regimen in all its parts, are the best auxiliaries which the surgeon can employ in the case of all injuries of the visera, contained within the cavity of the abdomen." With us in the East the state of our patients necessitated a much more cautious use of the lancet in these and in all other injuries, than is common. Opium, however, was the chief reliance in these lesions, as it allayed that pain and anxiety which might, without it, have been interpreted into a call for depletion. The most extreme abstinence from food is certainly one of the most important points in treating penetrating wounds of the abdomen. Purgatives by the mouth will do harm only, but clysters, especially of warm oil, are especially useful and agreeable to the patient.

---

* Hunter says of the blood passed by stool:—"If it is from a high part of an intestine, it will be mixed with fœces, and of a dark colour; if low as the colon, the blood will be less mixed and give the tinge of blood;" and of the character of the feeling, he adds—"the pain or sensation will be more or less acute according to the intestine wounded; more of the sickly pain the higher the intestine, and more of the acute, the lower." It would be a matter of some moment that we could rely on this sign. We can seldom, however, distinguish the character of the pain from the patient's statement, and it does not always afford us a true guide when it is recognized.

Few cases occur in military practice which demand the use of the suture to the intestine. Such cases are generally fatal. To those in which its employment is not distinctly indicated, Hunter's remark particularly applies:—" I should suppose the very best practice would be to be quiet, and do nothing except bleeding, which, in cases of wounded intestine, is seldom necessary."

Early protrusion of the gut is rare, unless the wound has been occasioned by a large ball, as a grape-shot. Its careful return is, of course, the rule of practice when it does occur. Guthrie has shown the propriety of leaving protruded omentum to act as a plug in the wound.

It is in wounds of the abdomen that the treatment by "debridement" retains its last footing. The fear of strangulation by the strong fasciæ, or between the muscles, is assigned as the claim it has to adoption in these wounds. But experience, while it has overthrown this cause of anxiety, has shown that a positive evil is occasioned by the practice, in so far as that the abdominal walls are weakened by it, and hernia the more apt to ensue. This step then is abandoned here, as in all other regions, unless an absolute necessity arise for its adoption. In the case of narrow wounds through the deep muscles of the back, by which fæces ooze, but cannot get a free escape, in similar wounds penetrating the bladder, or in cases in which a large amount of blood has been effused into the abdomen, it may be necessary to enlarge the wound, in order to prevent ulterior consequences of more gravity than those which can follow from the step itself.

If a false anus result from a penetrating wound by gun-shot, the cure will in most cases take place in time spontaneously. Of this I observed, with much interest, two cases at Constantinople, both of which very quickly got well. A plastic operation at a late date will probably supply what is deficient in the effort of nature.

Where the destruction of soft parts has been considerable, the danger of ventral protrusion will require attention during after-life, and no little trouble is often caused by the irregular action of the viscera, by pains which either wander throughout the cavity, or localise themselves at the point wounded. These uneasy sensations are increased by any distension, such as that which follows a full meal, and they continue to distress the patient during digestion. Dupuytren dwells on the effects of that chronic inflammation which may be set up by a contusion of the gut, and which, he says, may bring about a stricture of the intestinal canal, or its cancerous degeneration.

I had fewer cases of penetrating wounds of the abdomen under my notice in the East, than of almost any other serious injury. The following are given as among the most interesting of those of which I have retained notes:

Cousins, a private in the 77th foot, aged 18, was admitted into the general hospital, under Mr. Rooke, on the 8th of June. When standing in one of the advanced trenches sideways to the enemy, his right arm being stretched out in front of his hip, he was struck by a round shot or large piece of shell, which completely smashed his right forearm, and fractured the ileum of the same side, causing at the same time a lacerated wound of the right iliac region about 5 inches long by 3 broad. The wall of the abdomen, including the peritoneum, was destroyed to the extent mentioned, and a coil of intestine was laid bare. No protrusion took place, nor was the gut seemingly injured. Besides the fracture and destruction of the crest of the ileum, the anterior superior spinous process of that bone was quite detached, and the great trochanter was also fractured. The leg on the wounded side was shortened very considerably, and the foot was everted. As from the extent of the injury sustained and the collapse present, it was supposed that this patient would die shortly after admission, nothing was done for him beyond simply dressing his wounds and giving him stimulants in small quantities. Next day, however, he had so far rallied that some hopes were entertained for him, but it was not till the second day that he had sufficiently improved to allow of his arm being amputated. This was of course done under chloroform, otherwise it is questionable whether the operation could have been performed at all, the patient was so much depressed. He had at this time no abdominal uneasiness, and his bladder acted freely. By the attentive administration of mild nourishment and opiates, this patient gradually improved. No tenderness or other untoward symptom appeared in the abdomen. The wounds assumed a sloughy look for some days, and deep cellular inflammation in the upper part of the thigh made incisions necessary. On the fifth day, his bowels were for the first time moved by the aid of warm-water enemata. At this time the wounds were granulating kindly, and the stump was healing well. The coil of intestine was still visible at that date. The ala of the ileum, which had been laid bare, granulated over, but most of the crest became loose, and was removed at different times. The bowels came to act naturally, and without any stimulation, and by the end of July the wound on the abdomen had completely healed by granulation. The femur, if fractured (and of this there was every symptom, though the state of the pelvis prevented a careful examination being made,) became consolidated, but remained two inches shorter than the other. The simplest dressings, and almost no internal treatment, were followed throughout the progress of the case. This patient had never a bad symptom, but made a most excellent recovery; and when he went to England in September, all his wounds had healed with the exception of two small sinuses, leading to dead bone, on either side of the great trochanter. Below Poupart's ligament, and external to the emoral artery, a hard mass was traceable by the touch, which appeared to be some part of the pelvis driven down into that situation. It did not give him any annoyance. The limb, though shortened, was fully moveable at the hip-joint, without causing pain, and he could raise his knee, but not his heel, from the bed. The shape of the hip was destroyed, the projection of the crest of the ileum gone, but that of the great trochanter was unnaturally increased.

O'Neil, private in the 38th regiment, was admitted, under my charge, into the general hospital in June. A ball entered his left lumbar region, about three inches from the spine, as he was lying on the ground in one of the advanced trenches, with his feet towards the enemy's works. The ball lodged. The finger went deeply inwards and somewhat upwards, but detected nothing of the ball, the situation of which could by no means be made out. In the evening, his abdomen became a little tender, his pulse hard, and his face flushed. He was once bled, opium administered freely, and a fomentation applied to the belly. Next day the uneasiness had gone, and for eight days there was no return of it whatever. His alvine evacuations were, in the meantime, regulated by the use of mild clysters. No blood appeared by stool. The wound suppurated healthily. He was kept on very mild and easily-digested diet. On the eighth day severe pain suddenly set up in the left iliac region. This pain was increased by pressure, but was very limited in its extent. He vomited frequently, and his pulse rose to 110 per minute. His bowels had acted freely the day before. His tongue was dry and furred. He had a dozen leeches and repeated fomentations applied to the abdomen. Dover's powder, in doses of gr. x., was ordered every second hour. Next day the pain had quite left, and all treatment was stopped. His bowels did not act without the use of a clyster. He got plenty of mild nourishment, and, after a time, cod-liver oil. Though without any uneasiness or symptom of ailment, he became much emaciated, but ultimately rallied, and made a good recovery, the position of the ball never having been discovered, though the direction and depth of the wound would appear to favour the view that it had penetrated the cavity.

I saw a patient in one of the French hospitals at Constantinople whose abdomen had been traversed from behind, forwards, by a ball at Inkerman. The bullet had entered near the spine of the last dorsal vertebra, and had escaped near to, but slightly to the left of the umbilicus. The gut protruded for some days at the anterior wound, but did not appear to be injured, at least no intestinal secretion showed itself at either orifice. Hardly any bad sympton seemed to have followed. The gut was returned, the man kept low, and opium freely administered. He made a most excellent recovery. In another patient in the same hospital, a wound of exactly the same description had been inflicted. The same symptoms and result followed, except that the gut did not protrude, and that recovery was slower.

The following was a very remarkable case, which, though not strictly a wound of the abdomen, I mention here as I do not intend to refer to gun-shot wounds of the rectum. I saw the patient at Scutari, towards the end of 1854, under the immediate charge of Mr. Price, now assistant-surgeon of the 14th regiment. A ball entered the front of a soldier's left thigh, three inches above the patella, as he was mounting the heights at Alma, and passed upwards deep among the muscles of the thigh. It then turned round the limb, traversed the muscles of the left hip, crossed the perineum deeply, and escaped on the right hip, having passed through the rectum some way above the anus. The wound of exit closed, and for several days before death fæces passed by the wound above the knee. Sloughing and irritative fever set in, and he sank rapidly.

To prevent the infiltration of fæcal matter in these cases, Larrey has recommended the use of a tube in the rectum.

The bladder has been wounded by a gunshot several times during the past war, but the returns fail to tell us how often.

Balls at times pass through the pelvis, and yet spare the contents.* Thus, in one case, of which I have notes, it passed in by one sacro-ischiatic notch, and out by the other, without doing more mischief than contusing the rectum. When the bones of the pelvis are broken, the injury is very serious, from their deep position, neighbourhood to important vessels, and thick covering. Stromeyer has called attention to the great liability there is to pyœmia after such injuries. If the ball passes through the peritoneum, then the risk of violent inflammation is so great as to render the wound generally fatal.

The bladder may be wounded in many directions, but the passage of the ball in an oblique line from above downwards, and to either side, seems the most common course for it to take. Occasionally its superior fundus is opened by a ball passing across the abdomen from side to side, close above the symphysis pubis. The gravity of the wound will depend mainly on whether the peritoneum has been injured or not. If it has not been opened, then the prognosis will, in some measure, hang upon the empty or full condition of the viscus at the moment of penetration. If the direction of the wound permit of the infiltration of urine into the peritoneum, then the fatal issue will not be long delayed. These are the cases whose hopeless nature probably gave rise to

---

* In the case of a man wounded at Chillianwallah, a six-pound grape shot passed through the pelvis, and yet he survived four days.

the oft-quoted Hippocratic axiom, "cui persecta vesica lethale;" as gun-shot wounds, at any rate, implicating those parts of the viscus which are uncovered by serous membrane, are by no means so mortal as they were so long supposed. Dr. John Thomson saw in Belgium alone, fourteen cases in a fair way of recovery.

A ball may lodge either in the neighbourhood of the bladder or entering its cavity, remain there. This latter result will be most apt to occur when the bladder is full of urine, or the ball much spent at the moment of contact. In rare cases a ball, when very small, has been passed with the urine, and it has been known to escape by the formation and opening of an abscess in the perineum.

The urine may escape by the wound at once, or at a later period when the eschar separates from the wound; or it may not escape at all. It is seldom, however, that it fails to pass in some quantity at the time of injury. The swelling which takes place in the lips of the wound prevents in a great measure the flow of the secretion by the opening; but it is by no means always sufficient to do so, as we would be led to suppose from Larrey's statement. The urine may, and does at times escape by both wounds if the ball has passed out; but from the greater amount of bruising and swelling which takes place at that of entrance, it may fail to appear there, even although it be the more dependent, and flow only from the wound of exit. The early passing and retaining of an elastic catheter is a most important part of the treatment of these cases, as it prevents the urine, in traversing the canal of the wound, from becoming infiltrated among the divided tissues. Larrey, recognizing the existence of this danger only at the period of separation of the eschars, did not employ a catheter early, but was particular in its use at the period when he thought the accident referred to was most apt to occur. Moreover, the fact that the slough is by no means the barrier to infiltration which he supposed it to be, is now well recognized, as well as that the exact period when its separation is to be looked for, we know, cannot be relied on. The irritation and straining which the unevacuated urine occasions, may prematurely force off the slough, and allow the urine to become effused, and so the mischief may be done before we are ready to combat it. Unless the wound implicate the neck of the bladder, the presence of a gum catheter will create but little irritation, and should be enjoined from the moment of injury. The catheter had best be retained till the urine begins to flow by its side, as the formation of abscesses with their disagreeable and dangerous consequences are thus more safely guarded against.

Larrey, with the object of obviating infiltration and venous engorgement, had recourse to scarifications, so as to enlarge the wound, and prevent all retention of secretion in its track. This step will, however, be perfectly uncalled for, if the catheter be retained from an early period. Rest, low diet, mucilaginous drinks, enemata, it may be leeches, and fomentations, or hip baths, will comprise the rest of the treatment in the majority of cases. The employment of morphia suppositories will also be found, under certain circumstances, most useful. If any urine does escape into the tissues, its early evacuation will of course be necessary.

The posterior or lower wound commonly closes before the anterior; but neither ought to remain long open if the catheter be made to remove the urine so soon as it enters the bladder. If the part through which the ball has passed be deep, the external orifice of the wound may close before the rest of the track—a result which should be avoided.

The position of the bladder, its depth from the surface, it size internally, the want of correspondence which takes place between the external wound and that in its walls, from their contraction after the passage of the ball, make the extraction of a ball by the wound a matter of impossibility, without such an enlargement of the orifice as would be injurious.

If the ball remains in the bladder, it becomes a matter of moment to remove it. Balls, pieces of cloth or bone so introduced, form the nucleus of calculi; so that the sooner they are got quit of the better, provided the immediate irritation and inflammation caused by the wound have subsided. Many cases are now on record in which the bladder has been opened, and calculi, having balls as their nuclei, have been removed. Larrey operated successfully on the fourth day after the introduction of the ball, and mentions a case in which Langenbec succeeded in removing a similar body ten years after its introduction into the bladder. Morand operated twice. Deuiarquay mentions a case in which the nucleus was a piece of shell. Baudens successfully removed the ball by an incision above the pubis; Guthrie by the lateral operation. Hutin mentions two cases in which a ball or foreign body was removed by lateral incision, one after thirty-two years', and the other after nineteen years' residence in the bladder. In one of these cases three calculi were removed, having pieces of cloth as their nuclei. Besides these, Mr. Dixon, in the 33rd volume of the *Medico-Chirurgical Transactions* has given the particulars of ten other cases in which balls were successfully removed, and three in which the attempt failed. Nearly all of these patients were operated on years after being wounded. In the *Medical Examiner* for 1855 a case is recorded in which a large ball, driven into the bladder, was not found till two years after, on the death of the patient. It formed the centre of a large calculus concretion.

The following case I find detailed in the Report from the sanitary depot at Landour for 1849–50.* Private West was wounded on the hip by a grape shot at Chillianwallah. The ball was lost, and the wound healed kindly in six weeks. A day or two after being wounded, he experienced a scalding sensation in the urethra on micturating, and he showed marks of a urethral discharge on his linen, which he thought was a return of an old gonorrhœa. He was treated under this idea for a time, the symptoms of inflammation in the bladder being ascribed to the gonorrhœa. The attacks of cystitis became so severe as to cause his bladder to be examined, when a hard substance was discovered. The introduction of the instrument gave great pain, and it was only on the second trial that a foreign body was detected. By the lateral operation a grape shot was found and extracted, "slightly incrusted with a sandy deposit." He recovered perfectly. No bone was injured by the ball. "After the operation the patient remembered that he used to pass blood and pus in his fæces after he was wounded. Hence it is probable that the ball entered by the sciatic notch, and traversed the rectum, entering the bladder at its back part."

The following is a fair example of a penetrating wound of the bladder:—

Griffith, private 57th regiment, was admitted into the general hospital in the summer of 1855. A ball had entered his left hip, close to the tuber ischii, and escaped on the abdomen, two inches above the symphysis, a little to the right of the middle line. Urine escaped by the anterior opening. A catheter was passed into the bladder and retained there. He had no bad symptoms of any kind for twelve days. His urine passed by the catheter, and also by the opening on the abdomen. His pulse remained quiet, and his abdomen without uneasiness. His general health was unimpaired, and his bowels acted regularly. The posterior wound, through which urine never passed, closed rapidly. On the twelfth day, he had severe pain in the abdomen, which was, however, relieved by a dose of opium, and he never afterwards had a bad symptom or uneasy feeling, except the irritation occasioned by the urine flowing on the abdomen, which could not be altogether prevented. His urine was loaded with mucus and pus during the period of cure, and he passed several small pieces of bone, both by the urethra and by the abdominal wound. At the end of six weeks he could retain his urine, and pass it at pleasure by the natural passage, in a full stream. For a month he had been unable to prevent his urine flowing constantly away. In about two months from the period of his admission the wound

---

* Unpublished Records of Medical Department. This case is referred to by Guthrie, and has been recorded by Mr. M'Pherson, in connection with Mr. Dixon's paper, but with some variation from the account given in the text.

on the abdomen was completely closed by the use of nitrate of silver. His strength, which had somewhat failed, was at that time quite restored, and he was walking about the ward convalescent. At this period he passed from under my notice; but I learned that the wound on the abdomen had reopened, and that he could pass his urine, without any pain, through this opening, in a continuous stream, but that ultimately, before he went to England, it had permanently closed.

The following case is curious, as showing how large a body may descend into the pelvis, and yet very slightly injure the viscera. A soldier at the Alma was wounded by a piece of shell, which struck him over the symphysis pubis, and descending into the pelvis, was lost. No bad symptom whatever supervened, and he made a rapid recovery. The surgeon in charge of the case thought that the missile lay impacted deep in the pelvis, behind the pubes, but this he could not satisfactorily determine. Here the bladder escaped most miraculously.

The injury was much more severe, but the result little less fortunate, in the following case. A French soldier of the line was struck at the Alma by a piece of shell, above the symphysis pubis, which fractured the bones, passed downward, and was removed in the perineum from the side of the urethra. The rectum and urethra were both lacerated. Deep abscesses formed, the patient's strength gave way, but no acute attack of inflammation seized any of the viscera. A communication was established between the bladder and rectum, and between the bladder and the abdominal wall, so that gas and small pieces of fæces escaped at times on the abdomen. Blood frequently passed by the urethra. The last time I saw this man was in January, 1855, when he was recovering rapidly.

In the next case the missile penetrated the pelvis from below, and it is interesting chiefly from the manner in which the peritoneum escaped. A French artilleryman was wounded at the battle of the Alma by a piece of shell, which struck him on the perineum, and penetrated between the rectum and bladder, establishing a fistulous communication between these parts. The peritoneum was not opened. No bad symptom followed, but when he was sent home he was dying of phthisis.

There is a case related in one of the Indian reports, which illustrates in a curious way the severe injury which the perineum may undergo. A soldier of the 14th light dragoons had the pommel of his saddle struck by a round shot at Goojerat. The ball passed under and between him and his horse, which escaped injury. The rami of the ischium and pubes were fractured on the left side, the perineum extensively lacerated, but the scrotum was only slightly abraded, and the urethra was uninjured. He had much pain afterwards in passing his urine; the soft parts of the perineum sloughed, and his testicles atrophied; but otherwise he made a good recovery.

CHAPTER VII.

COMPOUND FRACTURES OF THE EXTREMITIES, GUN-SHOT INJURIES OF THE HAND AND FOOT.

In the returns of the late war, from April 1st, 1855, 2198 cases of gun-shot wounds of the lower extremities appear among the men, and 166 deaths therefrom. Of these, 1628 cases and 55 deaths were mere flesh wounds, and 43 cases and 2 deaths, wounds with contusion and partial fracture of long bones; 23 cases and 1 death, simple fracture of long bones by contusion of round shot; 174 cases and 64 deaths from compound fracture of the femur; 66 cases and 9 deaths from the same injury of the tibia or fibula alone; 144 cases and 27 deaths from compound fracture of both bones of the leg; 88 cases and 7 deaths from perforating or penetrating wounds of the tarsus. Besides those who died directly from the injury, 95 cases of compound fracture of the femur, and 91 cases of compound fracture of both bones of the leg were submitted to amputation.

There were 1237 cases and 8 deaths from flesh wounds of the upper extremity; 102 cases and 12 deaths from contusion and partial fracture of the long bones (including the clavicle and scapula); 27 cases and 2 deaths from round shot simple fractures; 169 cases, 15 deaths, and 104 submitted to amputation, from compound fracture of the humerus; 66 cases, 2 deaths, and 41 amputations from compound fracture of the bones of the forearm. In 113 cases the structures of the carpus were penetrated or perforated, and 48 of these cases were subjected to amputation.

Of all the severe injuries received in battle, none are of more frequent occurrence or of more serious consequence than compound fractures. They cause peculiar anxiety to the surgeon, from the manner in which their extent and gravity are so often masked, and from the uncertainty which still prevails as to many points in their treatment. This ambiguity as to their management arises in a great measure from the many varying causes connected with the state of health of the patient, and the means at hand for his treatment—circumstances which fluctuate with every campaign.

In the Crimea, these injuries were peculiarly embarrassing and extraordinarily fatal. In the management of no accidents was so much expected from modern improvements, and by none were we so much disappointed in the results. It was confidently hoped, that in very many of those cases which, in the old wars, would have been condemned to amputation, the limb would now be preserved, either by the exercise of greater care in the treatment, or by having recourse to some of the modern expedients by which limbs are so often saved at home. But, unfortunately, a sad experience only confirmed the hopeless nature of compound fractures of the thigh by gun-shot, and their very uncertain and dangerous character when the leg or arm were implicated.

In the following remarks on compound fracture, I propose to refer chiefly to those cases in which the femur was broken, and I will notice afterwards similar injuries of the leg and arm.

It can hardly be doubted, that the great striving after conservatism which influenced all the surgeons of our army, was one main cause of that mortality which attended these injuries. We were not prepared to believe how hopeless they were, till the unwelcome truth was forced upon us by an ever-recurring experience. We were disposed to judge of compound fractures by gun-shot as we would of accidents, similar at least in name, seen in civil life. Full of the promise of the schools, we would not admit that any injury apparently so slight could withstand the assiduities of a wise conservatism. In trying, however, to save limbs we lost many lives, thus fulfilling the prophecy of one of the greatest surgeons. Cases of promising appearance were reserved for the trial—the very cases, in fact, which would have made the best recoveries if operated upon early, and the inevitable amputation was delayed till the patient's constitution had become so depressed as to be beyond reaction.

Two circumstances seem to have had chiefly to do with the irreparable character and mortality of compound fractures of the thigh in the Crimea—first, the state of health of the men when wounded; and, secondly, the effect on bone of the new kind of ball with which most of these injuries were inflicted.

As to the state of health of our patients, it was not merely that they were in so anæmic a condition that suppuration and irritation quickly prostrated them ; nor was it that their stamina and "pluck" had been destroyed by hardship and suffering ; nor that the means of treating them in front, during the early period of the war, were totally wanting; but the chief cause of the reluctance shown by nature to repair the osseous breach was the scurvy-poison which held command in their systems. This it was which mainly opposed recovery. Callus was not thrown out at all; or if it was, it refused to consolidate. I myself examined the limbs of a large number of men who died at Scutari during the early part of the war, and in not a single instance almost did I observe the slightest attempts at repair ; but, on the contrary, invariably found a large sloughing chamber filled with dead and detached fragments of bone, shreds of sloughing muscle and destroyed tissue into which the black and lifeless bones projected their irregular extremities, and across which, lying in every direction, but seldom in the axis of the limb, were dead and detached sequestra, the "fracture-splinters" of the accident.

The depressed condition of body to which the hardships of the war had reduced the men, made a severe compound fracture of the femur synohymous with death; so that we might with perfect appropriateness use the words of Ravanton—"I exhausted many times the resources of art without success,—incisions, removal of the fragments, early bleedings of sufficient magnitude, spare diet, dressings, position, infinite care, nothing could protect them against an inevitable death." Most of our patients, as I before remarked, had either suffered from dysentery or were on the verge of falling into that disease. The vast majority of them had ulcerated intestines, and were thus in a condition of health which did not bear disease. When men in this state received a severe compound fracture, and their constitutions were taxed to repair the injury, there was no reserved fund on which to draw. They had been living up to their income of health, and so utter failure was the sure result of increased expenditure. If when injured they had been taken into the ward of a London hospital, I doubt whether they would in most cases have ended more fortunately, either by preserving the limb or by amputation; how much less, then, when they had to undergo treatment in a camp!

Many of our patients looked very well at first—appeared, perhaps, strong enough, and expressed such a confident hope in the result as almost to deceive their surgeon. The injury might not appear very severe; the bone was undoubtedly broken, but it might not be much comminuted; and thus we flattered ourselves, and began a trial hopefully which always ended in disappointment. The golden opportunity was allowed to pass, and so we entered on a road which led to death, whether through the portal of amputation or any other. The struggle soon began. Suppuration set in. The disease which lurked in "blood and bone" showed itself. Diarrhœa appeared and would not cease. The patient's stomach refused the only food which could be procured. He got emaciated, weak, and irritable. A suspicion was awakened that the bone had been more severely injured than was at first supposed. Things went on from bad to worse. Hectic claimed its share of the waning strength; and whether we operated late or not, the great regret remained that it was not done at first, as the invariable result demonstrated the uselessness of any other proceeding.

During the greater part of the siege, the means of treating these accidents, whether as regards food, bedding, clothes, or shelter, did not exist in camp; and to transfer them to the rear only made the fatal result the more certain, from the pyœmic poisoning which was sure to be set up by the transport. Thus then, it came to be, that up to the period when things were improved in the camp hospitals and in the transport service, recovery from a compound fracture of the thigh was impossible, or nearly so, and that the best hope lay in an early amputation. The only exception to this I will afterwards allude to.

bone with any touch more gentle than what occasions its utter destruction. In the Crimea we had many opportunities of observing the action of both kinds of ball, and so far as I could judge, their effects were so dissimilar, as almost to justify a classification of injuries founded on the kind of ball giving rise to them.* The longitudinal splitting of the bone is so dextrously and extensively accomplished by these balls that, while but a small opening may lead to the seat of fracture, the whole shaft may be rent from end to end. I have repeatedly seen the greater part of the femur so split. Stromeyer has shown that this longitudinal splitting seldom transgressed the line of the epiphysis, an observation which I can most decidedly confirm; for though the injury has at times been sufficiently severe to implicate both, yet the rule has been just as he says.

Gun-shot fractures of the long bones of the extremities have always been considered dangerous, chiefly on account of the shock, the comminution of bone, and the fact, that the wound leading to it is of such a character that it can heal only by suppuration, and cannot be so closed as to convert it into a simple fracture, which, it is well-known, we can sometimes accomplish in such fractures as present themselves to us in civil practice. The cavity of the fracture is thus kept open to the air; the pus undergoes those changes which Bonnet has shown it does under such circumstances, and that severe and prolonged inflammation of the deep and irritable tissues which constitutes the chief danger in compound fractures, cannot be avoided. Now, all of these dangerous characteristics of compound fractures have been immensely increased by the conical ball. First of all, the shock it occasions is undoubtedly greater than that caused by the round ball, simply because the destruction it causes is much more severe; secondly, the comminution of bone is enormously increased. The number of fragments which are quite detached are much more numerous, and the amount of sequestra, which are so far severed as to be ultimately thrown out before a cure can be looked for, is much greater. Thirdly, the bruising of the soft parts is more extensive, so that the suppuration is more prolonged, and the chances of purulent absorption so much the more multiplied.

The great loss of substance which follows compound fractures by the conical ball, is the source of one peculiarity in their treatment. The shortening will be greater should consolidation follow, than if the injury had been occasioned by the round ball.

---

* In these remarks I refer merely to the heavy conical ball, as there are balls of the same shape but of less weight, which are by no means so formidable. That used in the Schleswic-Holstein campaigns appears to have been very trivial in comparison to the large Russian one, of which we had such dire experience.

The conviction has been strongly impressed upon my mind, by the observation of not a few of these cases, that we ought not to keep up extension in their treatment, except in a very modified degree. If we do so—if we drag and haul at the bone, as I have often seen done, what is the result? A large hiatus exists, void of organizable material for forming the bone; the parts active in repair are drawn far apart, and a tax is made on the reparative process, which I will not go the length of denying may, under the most favourable circumstances, be brought about; but which I am fully certain never could be accomplished with us. In many cases it would, to my mind, be better practice—*i. e.*, it would afford better results in saving life and limb—rather to approximate than draw apart the fractured ends in such cases. Allow the ends of the bones to be drawn by the muscles towards one another, *having first removed the sequestra*, and attend merely to keeping the limb as straight as possible; or, in other words, do not be troubled with the displacement as to the length, but only as to the thickness of the bone, and I believe our chance of success would be improved. Deformity we would unquestionably have—shortening and twisting, and a limb of which I, for one, by no means recommend the keeping; but if we *must* save the extremity, if its retention is to be the test of good management, then I think our hope must be in some step like the foregoing.

There are rare instances of compound fracture which seldom present themselves now-a-days, in which the bone is but little comminuted, and which demand a different consideration altogether from those I have been speaking of. These accidents commonly arise from the contusion of a round shot, or the contact of a piece of shell. They are, however, so very rare and difficult to recognize, that less harm will follow from the same line of practice being pursued with them, viz., that of immediate amputation, than if by being careful about such rare exceptions, we run the extreme hazard of sacrificing the majority of cases which determine the rule.

The extensive comminution of the bone by a conical ball makes the indications with regard to the management of the sequestra more evident than it is commonly considered. I do not think we paid sufficient attention to their removal in the East. It may be true, as some tell us, that in fractures with the old ball, it was desirable to meddle as little as possible with the fragments; but this is the teaching of only a few. However, to my mind the question assumes a totally different light when viewed by the pathological results we had occasion to witness. It may be remarked, before proceeding further, that it is impossible not to recognize the practical nature of the division of the sequestra made by Dupuytren into primary, secondary, and tertiary, according to their degree of connection with the parts, and this, notwithstanding Esmarch's assertions to the contrary; nor can I see that the distinction of them, proposed by the latter, into "fracture-splinter," and "necrosed-splinter," makes the thing a whit clearer, or the division a bit more useful; so that in the following remarks I will adopt the old division.

The longitudinal sections into which the bone is split are mostly capable of consolidation, except at points where their connection, or the contusion they have undergone places such parts of them in the position of tertiary sequestra, which will exfoliate at some undetermined date. These fragments cannot of course be touched. The secondary splinters, again, or those loosely connected—hanging by an extremity, or by an edge, to the periosteum or to the tissues—are commonly very numerous, and lie by their detached parts in all directions to the axis of the shaft. The primary sequestra, or those wholly separated from their connection by the accident, are, in fractures from the conical ball, peculiarly numerous and destructive in their action. In some cases which I have had an opportunity of examining, these were found, not only at the seat of fracture placed in every possible position except the right one, but also driven deeply into the soft parts on the side of the limb next the wound of exit,—long sharp delicate chips, whose presence must have been the cause of continued suppuration, of low disorganizing inflammatory action in the soft tissues and bone, which extended its ravages to limits far beyond the seat of injury. In one case which I observed in camp, where partial consolidation had taken place, the dead sequestra had become so involved in the new bone, and were so prominent, so irregular, and so rough, as to look like the bristles of a porcupine. When to these considerations we add the chance of other foreign bodies, pieces of accoutrements or cloth, remaining between the broken fragments, and the ideas suggested by the very narrow opening to the surface which remains in gun-shot wounds; further reasons will be seen for the practice which, I believe, should be in general followed—namely, enlarging the *exit* wound (especially if it be the more dependent, or if it be a conical ball which has occasioned it,) extracting all loose and slightly-attached fragments, and keeping the aperture open, so as to allow of the free flow of the pus.

We have seen that the severe commotion at the seat of fracture occasions the formation of that large "foyer" which is found full of detached and dead sequestra, disorganized tissue, and acrid pus, and which, unless it be got rid of, continues to bathe the ends of the shaft, gives rise to inflammation in the medullary membrane, supplies a depôt of absorption for the uncollapsing veins of the bone, and finally causes constitutional poisoning. Now, as a ball traversing a limb carries the fragments it detaches towards its place of escape, it is evident that they will be the more easily got hold of and removed on that side of the limb. These are the grounds on which the practice, advocated above, is founded. Unless such a step as is indicated be had recourse to, I cannot see how it is possible, except in very rare and exceptional cases, to hope for the cure in the field of a compound fracture of a large bone by a conical ball. Dupuytren, recognizing the necessity of getting quit of these fragments, recommends the enlargement of both orifices to an extent so great as "that the fingers, introduced by either opening, should pass freely and meet without impediment." This, he thought, however, should be avoided, if the part was very thick and muscular. The proceeding sketched above, is in no way so severe as this, and would be probably as efficient in fulfilling the end in view.

All surgeons who have had much to do with gun-shot wounds, are agreed as to the propriety of removing those fragments which are wholly detached; but some oppose the removal of any which retain the least attachment. The objections which have been advanced against the extraction of these, are, chiefly, that they assist in the repair of the breach, by throwing out bone, and that if they do die, they will be extruded by the suppurative process. To this it is replied that, if these fragments are at all extensively attached, their removal is never contemplated, but if they are connected only by a border, or an end, to the shaft or the periosteum, they can contribute but very slightly to form callus, and will almost in every instance die. One small part that is covered by periosteum may generate callus, but the rest of their bulk will surely perish, and give rise to abscesses and fistulous openings; and the amount of irritation, constitutional disturbance, and wasting suppuration which they will cause, before they are thrown out by the eliminative force of nature, are such as to make it impossible for any but those whose constitutions are the strongest and most vigorous to withstand it. The length of time during which these spiculæ keep up the suppuration and retain the wounds open, not only render the patients the more subject to pyœmic poisoning, but what is of some consequence in military practice, detain the men longer in hospital; thus encumbering the wards, and keeping the patient longer exposed to an attack of those fatal forms of gangrene which prevail in such circumstances.

It is needless to quote authorities to show how practical experience has condemned the leaving of these secondary sequestra in the wound, as nearly all military surgeons are at one on the necessity for their removal. M. Begin thus formulizes his great experience in a communication to the Academy: "I do not know any precept more erroneous and more dangerous in surgery, than that which tells us to respect and retain the fragments of bone partly detached in fractures. These fragments almost never recover their vitality, nor become united to the body of the bone;" and he also tells us in another place, to remove not only "those pieces which are entirely detached, but also all those which are moveable, vacillating, and capable of being extracted without the necessity of too great destruction." M. Hutin, again, whose position in the Hotel des Invalides gives him larger opportunities of observing the effects of sequestra which have been left unextracted than perhaps any surgeon alive, says, referring to his recorded cases—" I have given several observations, *taken from among several hundreds*, in order to show that the portions of non-extracted bone end sooner or later by setting up eliminative action, which is always painful, often dangerous, and at times fatal. I have also reported other cases in which immediate extraction has been followed by positive cures, comparatively prompt. These instances confirm the principles stated above. Like them, or even more, they confirm this truth, that the secondary sequestra, if they are not hurtful at the time when the wound is received, or shortly afterwards, become so almost to a certainty at last. They demonstrate the necessity of removing them." Roux, Baudens, Dupuytren, Guthrie, and nearly all the leading surgeons who have seen many gun shot wounds, repeat the same thing. I had many times the opportunity of seeing that these partially detached fragments seldom lie in the axis of the limb; so that if they did come to enter into the new bone, they would be more a hindrance than an assistance to its assuming its functions, not only from their position, but also from their interposing between the principal sections of the fractured shaft, and preventing their contact and union. Their partially-necrosed condition makes them very liable to become separated by a future accident, and thus to be free to act more powerfully still as foreign bodies in the economy.

Finally, considering the question in all its bearings, it must appear pretty evident that the removal of fragments must tend immensely to simplify the wounds under consideration, and therefore, that not only should all spiculæ which are entirely detached be removed as soon as possible, but that the same line of practice should be followed with regard to those which are so far detached as to retain but slight connections, and whose continued vitality must be doubtful; that this step should be accomplished by enlarging the exit wound; and that the practice is especially necessary in those cases where the femur is implicated, and a conical ball is the wounding cause.

The tertiary fragments, or those extensively adherent, should of course never be interfered with. Parts of these fragments may subsequently exfoliate, but at what period this may occur it is impossible to say. They may not appear for months, or it may be for years. Mr. Curling has lately made the observation, that necrosed portions of bone in compound fractures are longer of getting loose when they are connected with the lower, than when attached to the upper part of the shaft.

Any operative interference thought necessary for the removal of sequestra should be had recourse to at once, before inflammation has come on, or otherwise it will be more difficult for the surgeon, and not only more painful, but also more hurtful to the patient.

The few attempts that I saw in the East to resect parts of the continuity of the femur, were certainly most unfortunate. Such a proceeding is manifestly much more severe and hazardous than that I have referred to above. The resections, however, did remarkably well in the leg and upper extremity.

In the classification of injuries which was followed in the Crimea, no distinction unfortunately was made between fractures in the upper, middle, and lower part of the femur, which prevents the discussion of several interesting points.

Although making every endeavour, I have only been able to find a record of three cases in which recovery followed a compound fracture in the upper third of the femur without amputation. In two of them the injury was occasioned by round balls, and the comminution was slight. In the third case I could not ascertain what species of ball had caused the injury. In one of these the patient, an officer of the 17th regiment, was in the highest health at the time when he was wounded (8th September), and was of a peculiarly buoyant and hopeful temperament. The ball entered behind, and was removed in front, a little below the great trochanter, by Dr. Ward of his regiment. This patient received an amount of attention which it would have been quite impossible to bestow in the field under ordinary circumstances. He had a mattrass constructed so that his wound could be dressed, and the bedpan introduced without disturbing his limb. He was wounded at a time when the comforts of camp-life were little behind those of home; and yet I have been informed, that although his limb was in a very good condition when he left for England, the trouble it has since given him, and the deformed condition in which it remains, make it by no means an agreeable appendage. Another case was that of a soldier of the 62nd, who was found a day or two after being wounded, lying in the dockyard stores of Sebastopol, under the charge ef the Russian surgeons. He was discovered when the place was evacuated, and carried to his regimental hospital, where he recovered. The fracture in this case was in the lower part of the upper third. It had been occasioned by a round ball, and the splintering was not great. This man, however, was in the best health when hit. He had just joined from England, and his injury was comparatively slight. The third man may be said to have had his limb consolidated, in so far as that a mass of callus was thrown out, which cemented the bone; but he died of purulent poisoning, and never left the Crimea. I could not find out whether it was a round or a conical ball which caused the fracture in this case. I know that the French had hardly any recoveries. One was, however, presented by the Baron Larrey to the Societié de Chirurgie last May. This officer had been wounded in the upper third, and the bone had been consolidated.* I never could hear

---

* The records of the Val de Grace do not say what sort of ball caused the fracture in this case.

of any other except a Russian, whose greatly shortened and deformed limb, I often examined at Constantinople. This man's thigh was quite firm, and had been allowed to unite almost without treatment. There were probably a few other cases, but they did not fall under my notice; although during constant wanderings through the hospitals in front and on the Bosphorus, I was unremitting in my inquiries after such cases. I am certain, however, that, although the instances of recoveries were rare, they were yet not so exceptional as recoveries after amputation at the same part, as will be afterwards more particularly dwelt upon; and thus it would appear that, so far as the experience of this war is concerned, we must conclude, that slight as the chance of saving life is in any case, it is still our part to attempt consolidation in preference to amputation, when the fracture is in the upper third of the bone. M. Simon of Geissen draws a like conclusion from a review of all the reported cases of the injury; but he extends the doctrine to the middle third, in which I cannot agree with him, for reasons which I will afterwards state. In the Schleswic-Holstien war they preferred amputation to preservation in such cases. M. Hutin, in the Invalides, was able to discover twenty-four cases of recovery after compound fracture by gun-shot above the middle of the thigh, but no case of recovery after amputation in the same part. This goes further to prove the position maintained above. In whatever way we decide, it is unfortunately too true that death will most commonly follow; but yet, when we do not operate, the patient may live in comparative comfort for several weeks, while, in the other case, he has to undergo a very fearful operation, and almost certainly dies within twenty days.

From the construction and limited range of the official returns, it is impossible to show in figures what was, however, a well-recognized result of the surgery of the war, that though union did in rare cases follow compound fractures in the middle and lower third of the thigh, still the ultimate per centage of loss was greatly less when primary amputation had been performed, than when limbs were saved, or tried to be preserved, or removed at a late period. When we take into consideration the fact so well brought out by the authors of the "Compendium de Chirurgie Pratique," that we should, on the one side, calculate those who die before the period for consecutive amputation comes round, as well as those who do not recover from it, and not merely those who die after being submitted to the operation—then the force of the teaching which inculcates primary amputation in these cases becomes much greater. Besides, as the cases which were retained for trial were always those in which the amount of injury was least severe, and the patients those most adapted for recovery, the presumption in favour of early amputation is the more decided. There can be little doubt that the chance of obtaining consolidation is greater in the lower than in the middle third, as is also the hope of recovery from amputation; so that, taking one thing with another, the experience of this war would lead to the conclusion, that when the thigh is fractured by a ball in the upper third, it should be saved, but that amputation should be immediately had recourse to in cases of a like injury occurring in the middle or lower thirds. Those fractures of a simple description, which at times present themselves, are not meant to be included in this remark, nor is it to be understood that, under more auspicious circumstances as to the condition of the patients and the means of treatment, better results than those we meet with may not follow the preserving of the limb. In fact, under ordinary circumstances, recourse should always be had to the steps I before spoke of, with regard to the removal of spiculæ in cases of fracture of the lower third, and then try to save the limb; but in a like injury of the middle third, the rule should be to amputate.*

It is certainly very much opposed to the modern ideas of conservatism to condemn limbs without a trial, and I am fully aware how difficult it is to become persuaded of its necessity; but the unwilling conversion at last is made, though it is generally gained by the loss of several lives. The French surgeons in the East fully acknowledged the hopelessness of these cases; but the fatality of amputation was, with them, little behind that of preservation. This experience is as old as the history of war, and comes repeated in renewed accents from every battle-field. Military surgeons are almost unanimous upon the necessity of amputating in the cases specified, and most civilians who have had an opportunity of seeing much of these accidents, have come to a like conclusion, as can be seen by the tenor of the communications to the Academy by the first surgeons of France. It would be mere waste of time to record the strong and decided verdicts which have been given on this point, and which find their summing up in the words of one of the greatest surgeons of any age or country, when Dupuytren says in one of his clinical lessons, "I have repeated it often, and I repeat it for the last time, after the facts which I have observed, chiefly in 1814, 1815, and 1830, that my opinion upon this point is unshaken. In compound fractures from gun-shot, in rejecting amputation *we lose more lives than we save limbs.*" The sagacious Hennen endorses the same view when he says, "I am well convinced the sum of human misery will be most materially lessened by permitting no ambiguous case to be subjected to the trial of preserving the limb." Larrey, Guthrie, and in fact all the leading military surgeons of modern

---

* The ambiguity in the foregoing sentence is found in the use of the expression "ordinary circumstances," by which it is evident that our author does not refer to what he subsequently calls "circumstances of war."

T

times, proclaim the same thing. That exceptions must sometimes be made is undoubted; but still they are only exceptions, and rare ones too. Cases of compound fracture near the knee peculiarly call for amputation, if the bone be split into the joint.*

The results which we obtained might most likely have been more satisfactory if the army had made another campaign. Our bad hygienic condition deprived us of the improvements made in surgery during the last half-century.

But, even in those exceptional cases which result in consolidation, the condition of the limb is not encouraging. To this Guthrie bears strong testimony from his experience after Toulouse. M. Ribes, as is well known, failed to find a single case of recovery either after compound fracture, or amputation in the middle of the femur, among 4,000 cases which he examined in the Invalides at the period of his first visit; but during subsequent years he saw seven cases there of "cured" compound fractures, five of whom died after many years of great suffering arising from the injury, and the other two he lost sight of, as they left the institution; but when last seen they were in a grievous plight, and he says, "it is probable that these two soldiers

---

* I cannot avoid giving the following remarks of M. Begin—"All military surgeons have begun by wishing to preserve. but as their experience increased, and their observation extended, they amputated more, and they gain the conviction that they are right. At the outset of my career I amputated less than I did towards the end of my service, as surgeon-in-chief of great establishments. There are certain cases, very often exaggerated, of wounded who pretend to have preserved limbs which the surgeon wished to remove; I have been present very often at the miserable death of persons who have refused the operation, or who, they thought, would avoid it. The small number of the first, who boasted loudly, cannot compensate for those much more numerous of the second, which caused me much sorrow. And besides, how often are these preserved members not a pitiable burden for those who carry them? Ask the surgeon of the Invalides if he is not asked every year by some of these old soldiers to deliver them from the parts which are an annoyance to them, and which cause them inconvenience and incessant pain. I think it is a great misfortune that our military surgeons should allow themselves to be seduced by some of the assertions which you have heard; this forgetfulness of the experience of their most illustrious predecessors, will cause certainly the loss of many men, which the art, exercised with a more reasonable energy. might save." "I know that there exist examples of recoveries with shortening, and fistulæ remaining for years," says Baudens, "but to save two with fractured femurs, and to heal them imperfectly, we will lose thirty, of whom fifteen or more would have survived immediate amputation."

died from the effects of their accidents, and if they did not, their condition must be greatly still more wretched." In all the seven cases there was union certainly, but it was attended by much deformity, necrosis, and caries. Long years of suffering, constant abscesses, exfoliations, atrophy, sensitiveness to the slightest atmospheric change, shortening and deformity. the development of phthisis if it be in the constitution—these are among the results of a "cure" of a compound fracture by gun-shot in the middle of the thigh.*

Finally then, let me repeat the conclusion—that under circumstances of war similar to those which occurred in the East, we ought to try to save compound comminuted fractures of the thigh when situated in the upper third; but that immediate amputation should be had recourse to in the case of a like accident occurring in the middle or lower third.

Many of the fractures of the leg were so severe as to call for early amputation. Severe shell or round shot wounds seldom leave much hope of saving the limb, but in a large number, however, of very unpromising cases, the leg was preserved. A great deal was done in the leg in the way of removing fragments. Guthrie says they can be extracted "to almost any extent and number," and he directs us, if necessary, to saw off irritating

---

* In the Punjab, and other Indian campaigns, I have been able to find the details of 24 cases of compound fracture of the thigh, (parts not specified,) in which the attempt at saving the limb was made. Of these, 14 died very soon; but of the ultimate state of the remaining 10, or whether they continue to survive, I find no notice. Dupuytren, in 1830, lost 7 out of 13 cases treated by him. Malgaigne, in 1848, lost 3 out of 5, all being select cases, and those not adapted for immediate amputation. Baudens, in one series of 60, which he mentions in his book, amputated 15 immediately, of whom 13 survived; 20 were amputated late, of whom only 4 recovered. The remaining 25, although tried "avec obstination" to be saved, all died miserably except two, who retained "a deformed member, unfit to fulfil its functions," and which, he says, they would willingly part with. "Taking a retrospective view," says Bell, "we are in true perspective all the dangers of a nine months' cure, which is but a weary travel, step by step, betwixt life and death. In this view we see the dangers of frequent fevers, wasting diarrhœas, foul and gleety sores; some dying suddenly of gangrene, some wasted by the profuse discharge and successive suppurations, new incisions, and unexpected discharges of spoiled bones; we see those who recover halting on limbs so deformed and cumbersome, that they are rather a burden than a help. In the very moment that we hear of such a cure, we know how much the patient must have suffered, and how poorly he has been cured; and we can, from the long sufferings of those who escape, tell but too truly how many must die."

parts of the ends of the shafts. If one bone only be broken, and the loss of substance in it is not great, the case will be the more promising, as the unbroken bone keeps the fractured one steady, and the soft parts in place. It is when a scale of the bone, however thin, remains, as we occasionally see in shell wounds, that the best results in the way of cure are obtained. Such was the case in a most successful instance of repair, in a man of the 20th regiment, under the care of my friend, Dr. Howard, of that regiment. I relate it because it may be looked on as an example of a class of cases which were not uncommon. A piece of shell struck the edge of the left tibia, and destroyed the greater part of the thickness of its shaft, from below the tubercle downwards for about three and a half inches. The fragments were removed at the time of the accident, or afterwards, as they became loose; the posterior shell of the bone being, however, entire, was carefully preserved. Four months afterwards this patient was sent to England with a strong and useful leg, whose only change was a slight bending outwards—a condition which generally remains in these instances. This case was just such a one as presents the best hope for a good result. I by no means would infer that some most excellent recoveries did not take place when resections were performed of pieces, including the whole thickness of the shaft of the tibia; but they were much more rare, and infinitely more tedious, than cases like the foregoing. When the leg is fractured low down near the ankle by a ball, the accident is much more grave than when it takes place at the middle of the limb. I have exceedingly seldom seen a case recover in which the tibia was split into the joint.

The free anastomosis which exists between the vessels of the upper extremity, the large supply of blood which they convey, the ready development of a compensating circulation, the less drain there is on the system during the period of suppuration, and the less call there is for the patient to retain a constrained and irksome position during cure, render many things practical in compound fractures of the upper extremity, which could never be attempted in like injuries of the lower limb. The injury indeed would need to be very extensive before we would think of performing amputation at an early period in gun-shot wounds of the arm; as, unless the vessels are destroyed, there are many most dreadful and hopeless-looking accidents from which the arm will recover; and besides, secondary amputations are so successful, and resections so often sufficient to fulfil the necessary indications, that primary amputation is never performed in the upper extremity, except under the most desperate circumstances.* Stromeyer recommends the trunk to be made

---

* The following is a curious instance of recovery from a most hopeless-looking injury. It is related in one of the Indian regimen-

the splint in treating these cases, so as to do away with all that
fear of motion in the fragments which exists, if they are treated
in the usual way. Unfortunately, however, as pus commonly
burrows and has so to be evacuated on the inner aspect of the
arm, it is difficult to carry such an idea into practice. Pirogoff,
it appears, was so displeased with the results of his attempts to
cure fractures of the upper extremity in the Caucasus, that he
was disposed to submit them all to amputation. The world will
learn with interest whether his experience in Sebastopol has not
been more favourable.

The results, with regard to fractures of the fore-arm, do not
tell the whole truth, as there is no provision made in the returns
for showing double injuries; many cases are made to appear as
having ended fatally from these and other comparatively trivial
injuries, which were in truth the result of a complication of accidents, of which this was the one chosen for registration. I have
known this occur often. Fractures of the fore-arm, when not
combined with other injuries, turned out most satisfactorily.
Hardly a case came under my notice which did not do well, even
although the comminution of the bones was very considerable.

As to the treatment of compound fractures, little remains to
be said beyond what has been already hinted at, or what is commonly pursued. Perfect fixture—a fixture so well secured as, if
possible, never to be disturbed during the process of consolidation; plenty of fresh air, the free discharge of pus obtained by
judicious and early incisions and by position, and *not* by manip-
ulations of the injured part; and the administration of tonics
and nourishment, but as little strong stimulation with brandy
and wine as possible—these comprise all the chief points in the
treatment.

Purulent absorption has been the cause of death in the vast
majority of those compound fractures which ended fatally. Pus,
occupying both the chief veins and the interstices of the bone,
was commonly found, and purulent deposits in the lungs very
generally existed. I do not think, looking at the question as a
whole, that our experience would lead us to subscribe to Velpeau's
doctrine, that " purulent absorption is more common among those
who undergo amputation, than among those who have severe suppurations, and preserve their limbs." Hectic, the renewal of old
enteric disease, and cholera, carried off many of our patients
under treatment for compound fracture.

---

tal reports in the War-office. A soldier received, in the Khyber
pass, a sword-cut, which divided his arm, bone and all, with the
exception of the vessels and nerves, and the muscles on the inner
side. He also received another wound, which laid bare the spine
and ribs; yet he recovered, the bone of the arm uniting. He died
afterwards of another accident. Two somewhat similar instances,
one from Percy, are related by Ballingall, pp. 343–4.

The results which followed the treatment of gun-shot wounds of the hand and foot, were very satisfactory in most instances. Balls perforating either, created a great deal of destruction, but the repair was not slow. "The talent of preserving" was well shown in the Crimean hospitals in these instances, and in general the results rewarded the endeavours made to save the member.

It is remarkable how few sequestra separate in gun-shot wounds of the hand, even when the shattering of the bones has been great. The extrusion of any large piece of bone seldom occurred, so far as I saw. In gun-shot wounds perforating the foot, the most marked feature was the great swelling which followed, and the extreme pain which this distention generally caused. How far the rapid cures obtained in the field may remain permanent, I am at a loss to know; but I fear that not a few of the cases "patched up," and sent home, may have to undergo operation at a subsequent date.

In dealing with gun-shot injuries, so severe as to demand operation in the field, we can often save more of the part of the hand or foot, than usually after accidents in civil life. The soft parts are seldom so much destroyed, in proportion to the injury inflicted on the hard tissues, by a musket ball as by a wheel of machinery; and thus we are not called upon to remove so much of the member in order to secure a good covering for the hard tissues.

CHAPTER VIII.

GUN SHOT WOUNDS OF JOINTS.—EXCISION OF JOINTS, ETC.

Gun-shot wounds of joints form a group of cases most interesting to the surgeon. "As for a wounded joint," says John Bell in his treatise on gun-shot wounds, "we may take the united experience of all surgeons, which has established this as the true prognostic, that *wounds* of the *joints are mortal*." Without, however, being so sweeping in the condemnation of such cases, it must be affirmed that no class of gun shot injuries prove more uncertain in their results, or are more commonly followed by disastrous consequences.

The gravity of gun-shot wounds of the joints will depend chiefly on the size and construction of the articulation, the extent of the injury, and the attention received by the patient shortly after being wounded—especially the means of treatment being at hand, and not necessitating long transport. As a very grave amount of destruction may be inflicted on the articulating extremities of the bones without much external appearance of such mischief, we are often deceived in our early examination of these cases, and this is one reason for delaying the adoption of decided measures, though delay so frequently proves fatal.

The wound of a ginglymoid articulation is, as a general rule, more severe than that of a ball-and-socket joint, chiefly from its more complex structure. Larrey noticed how often tetanus was caused by wounds of these joints, and every surgeon can testify to the extremely severe symptoms which follow their injury.

Although it is true in general, that a mere fissure extending into a joint may not be followed by serious results, still it is no less certain that even such apparently trivial accidents are often followed by the most disastrous consequences.

It is a matter of much moment to possess a decided opinion upon the treatment of gun-shot wounds of the joints, as in no class of cases is prompt action so much called for; and none in which, by the parade of a few successful cases, is the mind of the surgeon more apt to be misled. If, on seeing a case, we were able to decide what remedies were demanded for its management, then possibly much suffering and no few lives would be saved.

Gun-shot wounds of the neighbourhood of joints require much attention, not only from the fear of secondary implication of the articulation, but on account of the stiffness which is apt to ensue in it from long disuse during the period of cure. Artificial motion should be begun early in these cases.

The hip is too deeply placed, and too much protected by the surrounding parts and its own form, to be often penetrated by a ball; but when it is implicated, the destruction is commonly so great as to render operative interference in some form imperative. Alcock lost three out of four cases in which this accident occurred, and in the fourth case, "where recovery took place, the joint itself, there is some reason to suspect, was but remotely affected." Occasionally a round ball becomes impacted in the head of the femur, and may cause only a partial fracture of its neck. It is not easy in either of these accidents, however, to recognize the injury at first, as no sign of displacement or crepitation may be perceived. This is, however, rare; but the following is one case of this description. It is related in the register of the Depot Hospital at Colaba, in the archives of the medical department. Alexander M'Phail, aged 33, wounded at Dublin, 24th March, 1843, *by a matchlock ball*, which entered a little above the great trochanter of the right limb anteriorly, and was lost. His leg became powerless. On coming to Colaba on the 26th April, he did not complain of much pain, except when the joint was moved. Slight fullness over the hip was the only symptom of injury. Leeches and counter-irritation were employed, and he seemed to get better. On May 6th he was attacked with trismus, and died on the 9th. The ball was found embedded in the head of the femur, which, with half of the brim of the acetabulum, was shattered, and the capsular ligament formed the sac of an abscess which contained a considerable quantity of pus and spiculæ of bone. The orifice of the wound, it is added, had closed some time previous to death. Larrey mentions the case of an officer wounded in Egypt, who received a ball in the neck of the femur. The wound closed, and twenty years afterwards, on the death of the patient from disease of the chest, the ball was found impacted in the bone.

The knee when penetrated by gun-shot presents an injury of the gravest description. Taking much interest in cases of this description, I visited every one I could hear of in camp, and can aver that I have never met with one instance of recovery in which the joint was distinctly opened, and the bones forming it much injured by a ball, unless the limb was removed; yet the returns show several recoveries after such wounds, some of which, at any rate, I cannot but think are founded on error. I have conversed with many surgeons of large experience on the subject, but never heard of any case recovering without amputation, in which the diagnosis of fracture of the epiphysis was beyond doubt; yet such cases have been put on record. I remember one case, probably included among the recoveries, in which a ball passed near the joint, causing some effusion and swelling in it, with no constitutional disturbance whatever, and resulting in the man's return to duty within a fortnight, but which the surgeon in charge put down as a penetrating wound, remarking (as he well might) on the curious immunity from constitutional or severe local symptoms which had marked the case.

The following is a very interesting case, and certainly one of the most difficult to explain of any with which I am acquainted. I never saw the patient, the details have been kindly sent me by Deputy-inspector Taylor from Chatham:—"Private George Hayes, aged 31, 47th regiment, was wounded at the Alma by a grape shot, which entered on the outer side of the ligamentum patellæ, and passed upwards through the knee-joint, shattering the patella in its course, and making its exit at the anterior aspect of the thigh about its middle, partially fracturing it. The greater portion of the petella was removed in the course of treatment, as well as various fragments of the femur (exfoliations?); but firm union of the latter, as well as anchylosis of the joint, fortunately took place. At the time of his discharge he could sustain his weight upon the limb, and could walk about without crutches." I saw another case very similar to this at Scutari in 1855. In this instance, the ball had struck the man when he was about to kneel, and apparently fractured the head of the tibia. The ball was removed from the anterior part of the thigh. Scarcely any bad symptom followed, except that the joint swelled, was painful to the touch, and ended by losing part of its motion. If the articulation escaped the passage of the ball, the case was very curious.

The round ball sometimes penetrates the lower end of the femur or the head of the tibia without causing splintering, or opening the joint, or at least with an amount of injury to the capsule which is very light; and such cases may recover, and so shake our conclusions about others of a less anomalous character. Balls, too, may pass very close to the capsule and yet do it no harm, though these cases are put down as penetrating or perforating wounds of the joint.

It is undoubtedly often very difficult to know whether the joint has been opened or not, particularly if the ball is a small one, as was the case in one instance afterwards mentioned; and it very often occurs that the missile has run superficially under the integuments, or coursed round the bones, when it appears to have passed through the articulation. It is to be remembered, also, that the swelling of the joint may be merely the result of a bruise, or of the extension from the neighbourhood of the inflammation which has been caused there by injury, and is thus no sign of direct wound of the joint.

Another point which renders these injuries difficult of recognition when the bones are not much implicated, is the length of time which may intervene before the appearance of severe symptoms. A week may pass, and yet both the local and constitutional symptoms may be very slight. Sooner or later, however, the well-known signs of joint-injury are set up, sometimes with great rapidity and severity.

It is not difficult to understand the peculiar progress and fatal results of gun-shot wounds of the knee, when we consider how sensitive to injury are shut cavities when enclosed by such a delicate membrane as the synovial lining of the knee, and how feelingly such cavities resent the introduction of air within them; how rapidly they degenerate under the effects of this air; what a mass of closely-compacted tissues become implicated when disease is set up in such an articulation; how it is that bone, ligament, and soft parts participate in the injury; how wide the bony expanse is which enters into the formation of the joint; and what a large surface is presented for purulent absorption and transmitting inflammation, as well as how difficult it is for foreign bodies or morbid secretions to obtain free exit. These are the chief causes why the injuries under consideration are so often followed by dangerous and fatal results. In civil life, wounds opening the joint are commonly caused by cutting instruments. Foreign bodies are seldom introduced, and the bones entering into the articulation are little if it at all injured. The wound being carefully closed, often adheres, and, by appropriate treatment, little mischief may follow. But if a ball be the wounding agent, foreign bodies are almost sure to be introduced from without, or created within by the splinters. The ball's track must suppurate before it closes, and cannot be shut up and retained without the hazard of pus accumulating in the cavity; air thus gets admission, and works destruction. Foreign bodies cannot be extracted by so small an opening from a cavity of such a construction; and thus these gun shot wounds of the joint, though often apparently very trivial injuries, become the most serious almost of any which can be presented to us.

The primary dangers of these wounds are not great. It is in those which are set up afterwards that the chief hazard exists. The long and wasting suppuration, the tedious and dangerous abscesses, and the purulent poisoning are the principal sources of alarm. These abscesses are most curious occurrences in knee cases. They appear almost invariably among the muscles of the thigh; and while they may remain long unnoticed, they give rise to the utmost trouble and danger. They burrow along the bone, often stripping it of its covering, and yet are seldom apparently in connection with the joint. The escape of some small amount of the acrid secretions into the superficial or deep cellular membrane, sets up renewed inflammation and suppuration there, and thus abscesses form, whose connection with the original depot it is difficult to trace. These collections almost always occur in the thigh in preference to the leg. At a late period of the case, the joint puts on all the appearances of white swelling—and observation first made by Dr. John Thomson.

Military surgeons of all times have recognized the necessity of removing the limb early in these cases when the articulating ends of the bones have been fractured by a ball, and the experience of the late war fully bears out the practice. French and English surgeons were, I think, at one on this in the Crimea. In December, 1854, I saw upwards of forty cases in the French hospitals, and all died except those primarily amputated. I have heard incidentally of one case occurring in their army which recovered, but have failed to learn its details.* It is certainly very disheartening, as well as humiliating to professional pride, to think that we cannot save such cases without amputation. The very small amount of visible destruction which is so often present, the slight complaint of pain or appearance of disturbance which frequently exists at the period when the limb ought to be removed in order to insure success; the very pardonable unwillingness of the patient, especially if he be an officer, to submit to so dreadful an alternative, where there is, to him, so little apparent danger; all render difficult the adoption of those measures which a dire experience has shown to be necessary, for that amputation is our only resource all are agreed.

Guthrie has seen no case recover in which the limb was not removed. Larrey reports some, but they were instances of slight injury. Esmarch, from the fields of Schleswic-Holstein, says—"All gun-shot injuries of the knee-joint, in which the epiphysis of the femur or tibia has been affected, demand immediate amputation of the thigh. It is a rule of deplorable necessity already given by the best authorities, and which our experience fully confirms."

I have often contemplated the laying of the articulation freely open at an early period in these cases, so as to permit of the extraction of all foreign bodies, and the free escape of the pus which

---

* In the Indian Reports I have been able to find the particulars of nine cases in which the knee was penetrated, but the injury was apparently so light as to lead the attendants to try to save the limb. Every one died. Alcock has stated the proportion of cases in which the articulations are wounded, to other gun-shot wounds, as between 4 and 5 per cent., nearly one-half of which were of the knee. Of 65 cases in which an articulation was primarily affected, 33 recovered, 21 with loss of a limb, 32 died, 18 without amputation. "It is quite evident," he adds, "that if the 18 cases of death without amputation, and the 14 cases of subsequent amputations (assuming them to be unfavourable cases for treatment in the first instance), instead of being treated, had immediately been amputated, we should then have had for result, not a loss of 25, but of one-third, which is the loss from primary amputation. Two-thirds, therefore, or 16 out of the 25 would have been saved." Of 35 cases in which the knee was more or less implicated, 22 lost their lives, and of the remainder 8 lost their legs. "After such results, it is little to say that the five who recovered preserved good and useful limbs."

must afterwards be formed, the retention of which is undoubtedly one great source of danger. This might be attempted even although it were necessary to lay the whole front of the joint open by an incision similar to that for excision. The joint has been frequently widely laid open by cutting instruments, both primarily and for disease, and most satisfactory cures have been obtained.

If, however, the attempt is to be made to save the limb, the most rigid antiphlogistic treatment must be followed. Local bleeding by leeches, and the application of cold; the avoidance of all local remedies which are of a relaxing nature; the perfect fixture of the articulation, and the absence of all pressure; as well as the early evacuation, by free incision, of abscess, and of matter if it form within the joint;—these are the leading and evident indications to be followed. Hectic, with its common accompaniment diarrhœa, purulent absorption, with secondary implication of internal organs, and tetanus, are the causes which most commonly bring about a fatal result.

The presence of the articular cartilages would be of little moment, as they soon disappear; and if the bones were kept in close contact, firmly fixed, and all discharges allowed freely to escape, there is no reason why most favourable results might not be obtained. Such a step is in no way so severe as excision of the joint, and yet how successful has this been! The great sources of irritation and danger would be done away with, and if we had a healthy patient to deal with, I cannot see why we should fail.

If amputation is thought of, the sooner it is undertaken the better, as when the operation is performed late, after inflammation and suppuration have been for some time present, the results are very unfavourable. The joint, when opened, presents most of the characteristics of chronic disease, plus the immediate injury to the bone,—cartilages eroded, synovial membranes degenerated, and the products of inflammation effused into the cavity.

In those cases which are occasionally saved, the cure is very slow and very unsatisfactory. They occur only in those instances in which the bones have been slightly injured and the patients possessed of a first-rate unimpaired constitution, and when the means of treatment are of the most perfect description.

The particulars of cases in which the knee-joint has been penetrated by balls are so similar, that I will only give a few in detail. The first case was under my immediate notice, and from its presenting what might have been considered the most favourable features for conservatism, it gave me much interest. An officer of the 63d regiment, aged 19 years, was accidentally wounded by his own revolver during the reconnoisance in force at Kinburn, on the 21st October. He was half reclining on the ground at the time of the accident, his left leg stretched, and his right knee half-bent. His pistol was in his right hand, and close to his limb. The muzzle was directed downwards, and obliquely inwards, towards the middle line of the body. The ball, a small conical one, weighing four drachms, entered at the outer and superior surface of the lower third of the right thigh, about three inches above the border of the patella, and lodged. The wound appeared to lead into the cavity of the joint, and an indistinct grating was said to have been communicated to the hand on moving the patella from side to side. Some bloody greasy fluid escaped from the wound. On the patient's coming into camp, a few hours afterwards, I first saw him. There was then considerable swelling around the wound, but the motions of the joint were free, and unattended by almost any pain, and there was no swelling whatever of the articulation. There was one small spot over the head of the fibula, of which he greatly complained. Two tracks seemed to lead from the wound. One ran towards the inner side of the joint, and the other went along its external aspect; both were quite superficial. The position of the ball could by no means be determined. The patient's youth and strength, the absence of positive proof that the articulation had been opened, together with the possibility there existed of the ball having been deflected, and having passed down by the track along the external aspect of the limb, and lodged about the place so loudly complained of, in the neighbourhood of the head of the fibula, made us determine to wait. The most decided measures were immediately taken to ward off inflammation. The joint was fixed, and as he was taken on board ship, and put under the immediate charge of my friend Mr. White of the 3d Dragoon Guards, every care was bestowed on him. On the 24th there was some swelling of the joint, accompanied with pain. The wound of entrance was beginning to suppurate well. Synovia had not been seen to escape. His pulse was 78 and soft, and his secretions were natural. The penetration of the joint was believed by all, still an attempt to save the limb was determined on. The usual local and constitutional antiphlogistic remedies, including evaporating lotions, fomentations, leeches and cupping, antimony and calomel, were diligently put into requisition. By the 30th, the joint was much swollen, and very painful on pressure. A spot, about the size of a shilling, over the head of the tibia, was exquisitely tender. He was feverish at night, when the pain was always much exacerbated. An abscess formed among the muscles of the thigh, and continued to suppurate profusely after being opened. It had no apparent connection with the joint. Slight hectic set in. The pain in the joint became lancinating and throbbing to a most harassing degree, particularly over the head of the leg bones and over the patella. The articulation assumed all the appearance and feel of a joint affected with white swelling. This was the state of things by November 19th, when amputation was finally decided on, and performed in the middle of the thigh, the state of the limb not allowing of its performance lower down. On the 1st December he was transferred to the Castle hospital at

Balaclava, at which date the stump was suppurating kindly, but an erythematous blush overspread the integuments of the limb as high as the hip. This disappeared in a day or two, and the stump cicatrized, all except a small part in the centre. His strength seemed to improve. On the 4th, he had a slight rigor, which was repeated twice daily. He gradually sank, had cold sweats, dyspnœa, diarrhœa, and died on the 8th. The end of the stump was hollow, and contained much pus, and half an inch of the end of the femur was dead. The lungs were congested, but beyond this no particular appearance was observed. The vessels of the stump seemed healthy. In the removed limb, the tissues around the knee joint were found much engorged, the articular cartilages and ligaments were quite disorganized, and the cavity filled with turbid purulent fluid. The ball lay below the patella, in the intercondyloid notch. Its pressure on the end of the femur, lower surface of patella, and head of the tibia, was marked by complete erosion of the cartilages on these points. The bones were not otherwise injured. This case is highly interesting, not only on account of the difficulty found in detecting the ball, lying in the position it occupied, but also from the uncertainty which marked the line of procedure at the outset, the long absence of serious symptoms, the smallness of the ball, and the fatal result, notwithstanding the slight injury to the bones. If a free opening had been made early into the articulation, might we not have saved life and limb?

Miller, private, 31st regiment, was admitted into the general hospital on the 9th July. While his left leg was coming forward, as he was marching down to the trenches, he was struck by a piece of shell over the lower end of the femur, and the external surface of the knee joint. The wound was about four inches long, little lacerated, but deep, and opened the joint. The wound was carefully closed by suture, the limb fixed, and cold was applied. Inflammation was violent by the 16th, notwithstanding the employment of every means to moderate it. The wound opened and synovia escaped freely. The constitution did not apparently sympathise much for many days after the local inflammation was considerable. Pus was poured out freely by the wound, symptoms of pyœmia rapidly set in, and he ultimately died on the 3d of August. The external condyle was splintered into the joint, the cartilages were eroded, but little fluid existed in the articulation. If the joint had been freely laid open after the wound had failed to adhere, might not a better result have been reasonably looked for?

Shell wounds of the knee are, as a whole, not so dangerous as bullet wounds. They frequently merely cut the soft parts open; or if they injure the bone, the larger aperture which they leave acts beneficially in permitting the free discharge of secretions. I have known many shell wounds in the neighborhood of the joint, and not a few in which the articulation was opened and even the bones injured, ultimately do well, so far as saving the limb goes; more or less anchylosis following. The following short notes of three cases of this description were kindly sent me from Chatham, and are of more use as bearing on this point than other cases, the outlines of which I possess, as they record the state of the patient some time after being injured. Patrick Madden, private, 49th regiment, was struck by a piece of shell on the left knee, when in the trenches, on the 30th of May, 1855. The joint, if not opened, was very gravely injured. He was two months under treatment in the Crimea, with his articulation much swollen and inflamed, and when he reached England the wound had healed. He was invalided on account of "swelling, pain, and weakness of the knee," the joint being partially anchylosed, and its tissues thickened.

Private John Dwyer, 49th regiment, aged 29, was invalided on the 23d April, 1856, for "partial stiffness of the left knee," occasioned by a shell wound which partially fractured the patella.

Private James Callaghan, 95th regiment, aged 21, was invalided at the same station, on the 14th April, 1856, for "anchylosis of the left knee." This man had received a shell wound of the joint, but no mention is made of any fracture of the bones. The articulation was still painful at the period he was invalided.

A paymaster-sergeant, belonging to the 38th regiment, a dissipated nervous man, was admitted into the general hospital on the 18th of June. While kneeling, he was struck on the knee of the limb not on the ground, by a piece of shell, which was supposed to have lodged. The wound appeared to lead into the cavity of the joint, and much injury of the head of the tibia had been evidently produced. The patient would not consent to the removal of his limb, and being a non-commissioned officer, his desire was complied with. The limb was slightly bent, and laid on a pillow, while local and constitutional remedies were promptly applied. For eight or ten days no disturbance, local or constitutional, supervened. The joint then began to swell, became glazed and painful, and his stomach became irritable. The pain was chiefly confined to a point on the inner side of the joint. He went on, one day better and another worse, the joint always becoming more hopelessly diseased, till July 15th, when the limb was removed. For a time he did well, but ultimately sank in the beginning of August. The head of the tibia was much injured and split into the joint, while part of its shaft was driven upwards into the cavity, and indented the condyle of the femur. A piece of the head of the tibia was also driven downwards, and was impacted in its own shaft. The articulation was filled with a dirty pink-colored matter, but the cartilages were not diseased. The flaps had adhered to a considerable extent, but within them a large cavity was found filled with pus, decomposed tissue, and blood. This cavity extended up along the external surface of the bone to the trochanter major. This case was an example of a large class in which amputation was performed late for gun-shot wounds of the knee, and in which this large depot of pus had formed. These are cases in which adhesion of the flaps by the first intention should never be attempted, but the utmost facility given for the escape of matter.

I have seen only one case in which the patella being fractured by a ball, the joint was not at the same time opened. The bone was in that case "starred," but the ball did not lodge. The subsequent inflammation of the joint was slight, and the recovery good; the motion of the joint being, however, considerably interfered with.

Penetrating wounds of the ankle generally did well, although they required long treatment. This is opposed to the usual experience of such injuries, but very much seemed to depend on attention to two things—first, that the articulation was rendered perfectly immovable; and secondly, that one or other of the wounds was so enlarged as to allow the free escape of all discharges. If the original wounds were large they generally did best, as surgeons are unfortunately averse to render them free if they are not so originally. The truth of this remark I have had ample opportunity of verifying. Stromeyer is of opinion that if there is much destruction of the external malleolus we should remove the limb, as the foot takes on the appearance of valgus, and is useless. I have not observed this result.

The shoulder joint has recovered well in several cases which I have noticed, where a ball has passed through part of it, and even in cases in which a good deal of the head of the bone has been destroyed. I suspect, however, that the after consequences are not always so encouraging as the rapid healing would lead us to expect. I have under my charge, at the present moment, an officer who was wounded at the cavalry charge at Balaclava, by a rifle ball which shattered part of the scapula and the head of the humerus. Nothing was done in the way of extracting the broken fragments at the time of injury; and accordingly, besides the hazards of a very long and tedious suppuration, during which many considerable spiculæ have been extruded, and the formation and evacuation of long "fusces" of pus, he is yet subject to a constant recurrence of these purulent formations, and to the exfoliations of pieces of bone which seem to be set loose by the least over-exertion. His joint is quite anchylosed. If excision had been practised early, might not a more useful limb have been retained, and much annoyance avoided?

The much greater simplicity and superficial position of the shoulder, than of the other articulations, cause it both to suffer less and to be more manageable when injured. Balls sometimes pass very close to the capsule without opening it, or, at any rate, injure it but slightly. Of this, I believe I have seen several instances. Larrey has recorded a case in which a round shot passed across the shoulder joint; and although it only abraded the skin, it yet shattered, the head of the humerus, the scapular end of the clavicle, as well as the acromion and coracoid processes. This man was saved by the excision of the destroyed bone. He tells us he saved many cases in which the opening into the joint was not great.

If a ball remains impacted in the head of the bone, as it sometimes is known to do, then the sooner it is got rid of the better, as caries of the bone, disease of the joint, and either amputation or death will follow. One case occurred at Scutari, in which the ball was found after death firmly impacted in the round head of the bone. I find the report of a similar case as having occurred in one of the regiments serving in China in 1841-2. In this case, also, the ball was not removed, necrosis was caused, and the patient died of exhaustion on the fiftieth day. Malgaigne, however, reports a case in which a ball had been so englobed, and no disease whatever caused in the bone, where a considerable cavity contained the ball. Abscesses and fistulous tracks are the things most to be dreaded in all cases in which the shoulder joint is implicated in gun-shot wounds.

In the following case of penetration of the elbow, the distinction between a wound caused by a sword cut from one made by a ball was well shown. If a ball had passed across the articulation, fracturing the bones, excision would have been called for. A dragoon was cut across the elbow of his sword arm by a Russian horseman, at the heavy cavalry charge at Balaclava. The olecranon was completely detached, and the joint opened. The wound was immediately closed, the arm placed in an extended position, and cold employed to allay inflammation. Little more was done, and the divided surfaces quickly adhered, and an arm remained which, although not so free in its motions at that joint as it was formerly, was yet most useful, and would, I doubt not, become more so in time. Abscesses around the joint, and œdematous swellings of the hand and arm, are very apt to follow injuries of the elbow. Larrey thought gun-shot wounds of the elbow particularly dangerous, from the strong ligaments which surround it, and the little distensibility of the joint; and he recommends amputation when it has been largely opened, even by a cutting instrument, and blood has been effused into the cavity. I have seen several cases in which, after being traversed by a ball, attempts have been made to save the elbow without excising it, but such trials were anything but encouraging. The motion of the joint, and its consequent use, will be found much greater after excision than when the arm has been saved without such an operation. Dupuytren has pointed out how important in gun-shot wounds of the elbow the position of the aperture is. If at its inner aspect the secretions get easy exit when the limb is in its natural position, and thus the chance of a favorable result is greater; if, on the contrary, the orifice be on the outer aspect of the articulation, no position will allow of the free flow of the pus

7

except one, which will prove very fatiguing, and almost impossible to maintain for a long period.

The Returns show the following results as having been obtained from the resection of joints, from the 1st April, 1855, till the end of the war. The imperfect state of the official documents makes accuracy impossible with regard to the earlier part of the campaign:

Cases.

Head of femur............... 5, primary, of which 1 recovered.
"   "   .... 1, secondary, fatal.
Knee joint................... 1, secondary, fatal.
Os calcis, and part of astragalus, 1, recovered.
Os calcis alone.............. 1, recovered.
Head of humerus............. 8, primary cases, 1 death.*
"   "   .... 5, secondary, no death.
Do., and part of scapula...... 1, secondary case, followed by death.
Elbow joint.................13, primary, with 3 deaths.
"   "   .... 4, secondary, died from causes not connected with the operation.
Partial of elbow joint......... 3, no death.

The above lists by no means represent the whole number operated on. Those who underwent operation after the Alma and Inkerman, after the battle of Balaclava, and the first winter's work in the trenches, are all excluded, and thus a vast number of the operations of the early part of the war are omitted. In fact I cannot but think that in this way the *majority* of the operations do not appear, as the number performed after these early engagements must have exceeded those executed at a subsequent period.

The excisions of articulations injured by balls, although occasionally performed during the Peninsular war, never became a very general practice, nor was it applied to some of the joints which later years have shown its advantages in. The Schleswic-Holstein campaign was the first great war in which this conserv-

---

* Larrey performed excision of the shoulder in Egypt 10 times: 4 died—2 of scorbutus, 1 of hospital fever, and 1 of pest, after recovery. In 1795 Percy mentioned 19 cures after excision of the shoulder. Baudens had 13 recoveries from 14 operations (*Rev. Med. Chir.*, March, 1855.) Of 19 operations performed in Schleswic 7 were fatal, most of them from pyæmia. Legouest had 6 cases of primary resection of the shoulder in the hospital at Constantinople, of which 2 recovered. Thus, then, Hennen showed little discrimination in condemning the operation, when he says that it was "more imposing in the closet than applicable to the field."

ative proceeding was followed out on an extensive scale, and the results obtained by Langenbeck and Stromeyer attests its efficiency, although they appear to have had recourse to it in cases of very slight injury. These operations certainly mark the surgery of the age, as, in the words of Malgaigne, it may be said: "C'est une des plus heureuses tendances de la chirurgie de ce siècle, quand la necessité lui met le couteau à la main de ne lui concéder que ce quélle ne peut lui ravir, de sacrifier aussi peu, et de conserver autant que possible."

The Crimean war afforded a considerable number of cases adapted to the performance of resection, and I think our results will stand a fair comparison with others, when all the circumstances are taken into account. I will not say that excision was performed in as large a proportion of cases of injury of the joints as we were led to hope at the outset of the war it might be; but when a better acquaintance was had with the character of the wounds which this war presented to us, it was easy to understand how such should be the case. The shafts of the bones leading from the joints were often too extensively destroyed to enable the injured parts to be removed by excision; in fact the shafts were so often split, and their periosteal and medullary membranes destroyed, that the resection of the articulation did not suffice to save the limb. Surgeons soon recognized this; but yet it was by no means always easy to determine the true state of things about the joint, till the incisions necessary for resection laid bare the bones, and forced the reluctant operator to convert his operation into one of amputation.

The great success which has attended the excision of joints in civil practice, and a consideration of the fact, that the cases which fall to be operated on in the field are free, not only of local affections in the articulation, but also of any active constitutional disease, made us all naturally sanguine of obtaining the best results from such operations in the field; but unfortunately, the circumstances above referred to interfered so as frequently to leave us no alternative but amputation.

So far as my observation went, primary excisions were much more successful than those done at a late date, and this fact is evidenced, both as regards the final results and the length of the period of convalescence, so far as we have returns.

One advantage, with regard to the opportunity of performing resections early is, that we can in general tell at first sight that such an operation *at least* is called for. It is not so with regard to a large number of cases for amputation, but generally we can see at once that excision, at any rate, must be performed, although we may not be able to determine that subsequent amputation will not be required.

In the following observations I have not referred to the amount of mobility retained by the joint after our excisions, as the patients went from under my notice too soon for my being able to do so. It is certainly among the disadvantages of military practice during war, that one can seldom trace cases to a conclusion; not only so, but, in such cases as those of resected joints, very much of the after result depends on that careful attention which no one can render so scrupulously as the operator himself; and as it is never in his power to bestow this, we are not likely to have such favourable results as when in civil practice, the patient remains, till finally cured, under the same hand.

The shoulder joint is certainly that to which resection is most peculiarly applicable, from its superficial position, and simple construction. Interference with this articulation is therefore less disturbing to the constitution of the patient, and the results of the operation are more satisfactory, than those which follow a like interference with any of the other large joints.

The experience of the surgeons in Schleswic-Holstein led them to conclude that secondary operation is less unfavourable when the shoulder joint is implicated, than when a late operation is performed on any other joint. The only secondary excisions of the shoulder joint which appear in the Crimean returns, seem to have been successful.

I know of a few cases in which what may be termed partial resections of the shoulder were performed, i. e., cases in which less or more of the head of the humerus was extracted, without the whole being removed; and I believe the result of such cases to have been, on the whole, satisfactory, so far at least as the healing of the wound was concerned. It is most curious how much can be done in this way. The after mobility of the articulation will, however, be more restricted in these partial excisions than if the whole joint is resected, and thus an entirely new joint formation permitted. The instances were not, however, very numerous in which the destruction of the bone was so limited as to allow of this partial resection. Very much depends on the careful management of these cases afterwards, especially in guarding against inflammation.

In the old war they restricted excision to those cases in which the injury was confined to the head of the bone; holding that, when the shaft was much implicated, exarticulation should be preferred, and this was very much the doctrine acted on during the Crimean war. Guthrie thought the insertion of the deltoid the lowest point at which the bone could be divided with any prospect of success; but Esmarch has shown that as much as four and a half inches may be removed from the humerus, and yet a most useful arm remain. The ligamentous matter necessary to produce such a favourable result, requires a "plasticité" of constitute which our patients did not possess.

The fact, so clearly brought out by Stromeyer, should be always borne in mind in determining on operations at the shoulder joint, that in comminution of the shaft of a long bone, the fissures never extend into the epiphysis; in the same manner, injuries of the epiphysis only in extremely rare cases extend into the shaft, unless the bullet strikes the adjoining borders of both parts, in which case both are usually more or less seriously comminuted.

As to the best method of proceeding for resecting the head of the humerus, some little difference exists, as is the case with regard to the excision of some of the other joints also, between those methods adapted for military and civil practice. This arises, in a great measure, from the character of the injuries necessitating the operation in either case. The soft parts suffer little, and the bone is not diseased, although broken, in cases operated on in the field, nor are the parts bound together, as they so often are in the excisions performed in civil hospitals; hence, it follows that we can often remove all that is necessary through a much more limited incision of the soft parts, than we could if disease was the cause of the operation, or yet many of the accidents which occur in civil life, in which, although the joint may be but slightly implicated, the soft parts are yet so often greatly destroyed. In the case of gun-shot wounds too, the periosteum, as well as the ligamentous and muscular tissues, of the articulation can be retained, and thus a very great advantage is secured, according to the views of this operation set forth by Stromeyer and Baudens. A single straight incision will thus then very often suffice in resecting either the shoulder or elbow joint, and even in similar operations on the knee and hip, so that the maxim of Desault, that "the simplicity of an operation is the measure of its perfection," is perhaps better exemplified in military than in civil practice. However, as in gun-shot wounds two apertures commonly exist, and as it is desirable to include them if possible in the incision, we have a further illustration of the saying, that the surgeon should be bound to no particular form of operation, but should adapt his proceedings to his case.

One of the chief dangers following resection of the shoulder is the formation of sinuses and abscesses in the neighbourhood. The best mode of avoiding this, is to arrange the line of incision so as to give free exit to the pus. Stromeyer's semicircular incision over the posterior surface of the articulation fulfils this end better perhaps than any other. The joint is there very easily got at. Langenbeck's one straight incision on the interior aspect of the articulation, with or without the transverse cut suggested by Franke, gave much satisfaction in Schleswic, where it was largely put in practice; and I have myself seen most admirable results got by the straight incision of White through the deltoid: but Stromeyer's allows of the more rapid discharge of all secretions than any of the others. Baudens, as is well known, prefers a straight incision on the inside of the joint in front, and from his large practice of it in Africa, was highly pleased with its efficiency, believing that it best allows of that ginglymoid joint being formed, which, he says, takes the place of the former articulation. Whichever of these methods of operation we adopt—and they represent those which have received the preference from military surgeons during late wars—it seems conceded by them all, that we do not require in field practice such extensive incisions as we do in civil life, and that by such limited incision the muscular and tendinous parts can be more respected, and thus the hope of restored action be much increased. The report of Esmarch on the practice of Stromeyer and Franke shows us, that to cut across the fibres of the deltoid does not much interfere with its after usefulness, "as the upper edge applied itself to, and united with the articular surface of the scapula, and was thus fully attached and able to raise the arm. The healing was also quicker, as the space to be filled by granulation was much diminished in size by the application of the muscle to the glenoid fossa."

When the neck of the bone is broken here, as in the hip, so that it is difficult to seize the round head of the bone, the powerful forceps used by Mr. Ferguson in excision of the jaw will be found to do good service.

Resections of the elbow joint were more numerous in our army than among the French, yet the number of cases adapted for it in either force was but small. The numbers mentioned in the returns by no means, however, include all the operations of this kind which were performed.

The formation of the elbow make gun-shot injuries of it much more serious than those which implicate the shoulder. Larrey was particularly gloomy in his prognosis of wounds of the elbow, and reports many disasters from them. That resection of the elbow is much less fatal than amputation, does not call for proof now-a-days, as it has been a long-established fact. The question now is more as to the extent of the articular ends of the bones which can be removed, consistently with retaining a useful joint. My notes of cases occurring in the Crimea, unfortunately, do not enable me to throw any light on these interesting points. In Schleswic, out of forty excisions, only six patients died, and two others were unsuccessful; but in thirty-two instances the effect was very good. "As regards two of them," says Esmarch, "I have not been able to learn anything with reference to the power of motion they possess; of the rest, eight have been very extensive, nine more or less complete, power of motion; it is to be hoped of many of the remainder, that they will be able to obtain much increased mobility by means of zealous exercise of the arm. On the other hand, thirteen of the cases have a more or less complete anchylosis of the joint."

Several of the cases operated on in the East had undergone injury of all the bones entering into the joint, but no case came under my notice in which so much as four or five inches were removed, as was done in the war in the Duchies.

Partial resection—of which there were a good many cases—

did not, I think, turn out, on the whole, at all so well as complete ones. They were more tedious, more liable to fail, and less satisfactory when they succeeded, than when the whole articulation was removed. The following were cases of partial removal of the articulation:—

A soldier of the fusiliers had the head of the radius and a small portion of the lesser sigmoid notch moved, shortly after injury. Amputation had to be performed two months afterwards, there never having been any attempt made to heal by the parts. A soldier of the 23d regiment, admitted into the general hospital in camp, after the assault on the 7th September, had the external condyle, the eminentia capitata, and part of the trochlea, destroyed and removed, the soft parts being little injured. Sloughing set in, great constitutional disturbance followed, and amputation had to be performed three months afterwards. If complete excision had been early performed in both these cases, I believe we might have obtained much more happy results. These are only examples of several similar cases. But, on the other hand, such cases as the following have occurred. A soldier, when mounting the heights of Alma, was struck by a rifle ball, which passed across his elbow posteriorly, fracturing the heads of the radius and ulna, but leaving the humerus entire. The broken fragments were removed, and the humerus left untouched; and, after three months' careful treatment, this patient was discharged with a famous joint, which admitted of a considerable latitude of motion, and with which he could sustain no small weight. In the 9th regiment, a man was struck by a ball, which destroyed the inner condyle of the humerus, without injuring the ulnar nerve. The broken fragments were removed by Mr. Thornton, surgeon of the regiment. The subsequent inflammation was commanded, and an arm was retained, which came ultimately to possess three-fourths of its original motion. Esmarch thinks that "the extensive severing of the ligamentous apparatus of the joint is what deprives the wound of its danger, and that the less there is removed from the joint ends of the bones, the greater is the probability of anchylosis."

The complete fixture of the joint during the early period of treatment, as so strongly dwelt upon by Stromeyer, its constant support by a splint, even when being dressed; the elevation of it, so as to prevent œdema; its flexure at an angle of 130° to 140°, are all points of importance, both as regards the comfort of the patient, and the after results. Early passive motion before the wound is wholly cicatrized, but at once abandoned if any irritation or signs of inflammation appear, are also indications which late experience has stamped the value of.

The hip was resected six times; five being primary and one a secondary operation. One of the primary alone succeeded. Such success, although small, is yet encouraging, as compared with the results obtained from amputation at the hip, for which operation the excisions were substituted; for, as will be afterwards seen, out of at least 10 amputations at this joint in our army, and some 13 among the French, none survived, and in our cases at any rate the fatal result very rapidly followed the operation.

I performed the first operation of excision of the hip undertaken in the East on the 6th July, 1855, on a rifleman, whose case is subjoined:—

Couch, a soldier of the rifle brigade, was struck, on the 18th of June, by a ball, close below the elbow-joint of his left arm. The ulna was fractured by the bullet, which then struck the femur on the great trochanter of the same side. The trochanter and neck of the bone was split, and otherwise severely injured. The patient did not come under my care till the 5th of July, when I found a large ragged wound over the injured trochanter, from which a very profuse discharge of pus flowed. At the bottom of this wound the bone was seen to be hollowed out into a large cavity, and to be split in all directions. The bone was black and dead. The limb was not shortened or distorted. The wound on his arm gave him much annoyance, and the pain from the hip was so great, that he urgently requested some operation to be performed, which might relieve him. He was brought under the influence of chloroform, in order that the injury might be more completely examined than could otherwise be accomplished from the patient's irritable condition, and also to enable me to get the limb put into a proper apparatus. On enlarging the external wound, so as to make it dependent, and to allow the necessary steps to be taken for the removal of the dead portions of bone, a large fragment of the external part of the femur, which comprised what remained of the great trochanter, was found detached, and a fissure running upwards, apparently into the capsule. It was found impossible to remove the dead bone without opening the joint; and, as but a very thin shell of the shaft was sound, a consultation decided on the propriety of excising the head of the bone, and removing along with it what osseous substance was destroyed. This was done without difficulty, the original wound being increased a little upwards. Hardly a drop of blood was lost. The wound was lightly dressed, and the limb fixed on an inclined plane, so arranged that the large dependent opening retained, could be got at without having to move the patient. The relief from pain and irritation which was experienced almost immediately, was very marked and gratifying. Next day the patient's pulse was firmer, his expression very markedly better, and he declared himself as perfectly at ease. The limb, in a few days, was shortened about two inches. Suppuration became established; his strength improved; hectic disappeared: he slept well; and his appetite, which, before the operation, was nearly gone, was now restored, and he was totally free from pain or uneasiness. His pulse, on an average continued about 85 beats in the minute, and was of good character. He continued to progress most favourably for about a week. Part of the wound closed, and the rest of it was clean and healthy. At the end of that period he was suddenly seized with violent diarrhœa, accompanied by vomiting and severe cramps, and followed by suppression of urine, which continued for 18 hours previous to death. His stools soon assumed the characteristic appearance of cholera evacuations, his strength sunk, he became rapidly collapsed, cold, and blue, and died during the night. Cholera was prevalent in the camp at the time.

After death, some crude tubercles were found in both lungs. There was no symptom of purulent absorption anywhere. The vessels in the neighbourhood of the wound were healthy. There were old ulcerations in the intestine, and recent enlargement of the solitary glands. The left ulna was fractured obliquely up towards the elbow joint. The wound over the hip was sloughy, an action it rapidly took on shortly previous to death, and the cut end of the bone was smooth and unchanged. If I had seen this patient earlier, when the nature of the injury could have been more exactly determined, I would have contented myself with merely gouging out the destroyed portion of bone, trusting to the remaining scale to throw out callus, fixing the limb carefully, giving free exit to the suppuration, and strenuously supporting the patient's strength. The state of the lungs and intestine, as revealed on *post mortem* examination, made this patient, at best, but an unpromising subject for operation; but if the fatal disease which terminated his existence, had not supervened, I would have been sanguine of the result. When he came under my care, I feel sure he was in such a condition, if no operation had been undertaken to relieve him from the mass of dead bone which his system was futilely trying to get rid of, and which was setting up further disease all around it, he would, in a very few days, have died, exhausted by suffering. These are, however, points more easily settled on paper than at the bedside.

Mr. Blenkins of the Guards, operated on the next case, and has been good enough to send me the following notes of it:—

"Private Charles Monsterey, aged 24, third battalion Grenadier Guards. Brought from the trenches at midnight with a severe shell wound on the outer side of the right thigh. Examination showed the thigh-bone to be extensively fractured at the upper part, in the situation of the trochanters and neck, the fragments were much comminuted, and the surrounding muscles greatly lacerated. It was at once recognized as an appropriate case for excision, and the operation was performed half an hour after his arrival in camp. The wound was extended in a longitudinal direction, to the extent nearly of five inches, and the shaft of the femur sawn through at the junction of the upper fifth with the rest of the shaft. The muscles were next detached from the trochanter, and the capsule lastly divided. It was intended at first to preserve the head of the bone in the socket, but the capsule was so extensively lacerated, and the cavity being filled with blood, it was resolved to remove it. Very little blood was lost during the operation. Examination afterwards of the excised bone showed it to be fractured in fourteen pieces. The trochanter minor informed three, trochanter major three, shaft five, the neck three, besides numerous smaller fragments. The case continued to do well for the first three weeks; healthy granulations sprang up, both from the end of the divided shaft, and the surrounding cavity and acetabulum. At this period pain and swelling of the knee joint of the same limb supervened, the capsule of that joint became filled with purulent matter, the cartilages eroded, and he sank gradually, worn out with hectic symptoms, at the end of fifth week, in spite of every effort to support him. The case was doing remarkably well, and I had every hope of his recovery, until empyema came on."

Staff-surgeon Crerar operated on the third case, a private of the first battalion of the Royals, who was wounded in the Greenhill trenches at mid-day on the 6th of August. The wound, in this case, was slightly posterior to the great trochanter, and was not larger than a shilling. It had been caused by a piece of shell, which before it entered his thigh, had first struck a water canteen that hung by his hip. A comminuted fracture high up was clearly ascertained; but its exact position or extent was not defined previous to operation, although it was supposed to implicate the head and neck of the bone. "The trochanter was found broken into several portions, detached and imbedded in the contused muscles around, from which they were at once removed. The fracture was found to extend obliquely inwards, about an inch and a half along the shaft of the bone. The femur was now protruded through the wound, and I sawed off the whole of the fractured bone, leaving a smooth clean surface; I then proceeded to disarticulate the head of the femur, which was effected without difficulty. Scarcely three ounces of blood were lost, and little or no shock was induced." This patient was seized with rigors, and died of exhaustion on the 21st, *i. e.*, on the fifteenth day from being wounded. The internal viscera do not seem to have been examined, but as to the state of the femur, Dr. Crerar says "nature had not made the slightest attempt to repair the loss."

The next case, which was the only one where success followed the operation, was that of private Thomas Mackenena of the 68th Light Infantry, operated on by Mr. O'Leary, surgeon of that regiment. The age of the patient was 25, and he was wounded on the 19th of August, by a fragment of shell which struck him over the great trochanter and fractured it. It was thought that the fracture ran into the joint, a supposition which was disproved at the operation, as the head of the bone was uninjured. Five inches were in all removed. After operation, the limb was slung to the beam of the hut. This patient recovered in three months. I have lately heard of this man through the kindness of Lieut.-

Col. Stuart, commanding the pensioners in the Newry district. Dr. Shaw, who signs the report, states "the limb is two inches shorter than the corresponding one, and also considerably smaller; extension can be carried on partially, but he cannot flex the limb upon the thigh without placing his hand on the glutei muscles of the diseased side. Rotation, inwards and outwards, can be performed only to a limited extent. The wound over the joint is quite healed. The man's general health is good, but he cannot walk without the assistance of crutches."

Dr. Hyde operated on another case after the taking of the fortress. His patient, a private of the 41st had the neck of the bone severely comminuted by a grape shot, and died on the fifth day after operation. The cause of death is not given, nor can I discover it from the report of the case.

Dr. Combe of the Royal Artillery performed the sixth and last operation, on account of a gun-shot wound of the neck of the femur, in which, however, the head of the bone was not implicated. This operation was not'a primary one, but the patient survived a fortnight, and died of exhaustion; the most marked feature in the case being, that the pulse remained very high—never below 120—during the period he lived, while his aspect was calm, and such as "might have led one to expect a more subdued state of the circulation."

Death thus followed in 2 (1?) from exhaustion, in 1 (2?) from pyœmia, in 1 from cholera, and in 1 from a cause that is unknown.

It is a remarkable fact, that in these cases the head of the bone so often escaped, when the upper part of the shaft was fractured in pieces, which is probably to be accounted for by its protected position, and perhaps by the same cause as that before mentioned, with reference to the head of the humerus, viz., the non-extension of fracture of the shaft to the epiphyses.

Boyer and others have dwelt upon the depth of the parts, the strong ligaments, the difficulty of turning out the head of the bone, &c., as insuperable objections to this operation; but actual experience—both in civil practice, where it has recently been so often performed for disease, and especially in our attempts in the Crimea, where the soft parts were in their natural condition, and the difficulty of turning out the head of the bone increased by the broken state of the shaft—proves that there is no such barriers to its easy execution. The greatest difficulty lies in the after treatment. It is desirable to fix the parts thoroughly, and at the same time to allow of some change of position. Mr. O'Leary managed this, to some extent, by means of a canvas sling for the limb. The fixture cannot, however, be satisfactorily accomplished in this way, whatever power it gives of changing the position of the patient. I adopted the inclined plane in preference to the long splint, because I believe the position to be an easier one for such a case, and also because it permitted the free discharge of the pus, and the easy dressing of the wound without disturbing the patient. If the idea lately suggested at King's College, of slinging the whole body, could be carried out, it might afford many advantages in the management of excisions of the hip. As to keeping the limb in a good position during cure, I fear more important ends are lost sight of in striving after it. The uneasiness and irritation which the splints and rollers give, do much to prevent success. It matters little what the resulting length of the limb proves to be, if the patient's life is saved; nor does it greatly matter that it be somewhat out of the right axis.

As to the comparative advantages of amputation and excision at the hip in cases of compound fractures of the head and neck of the femur by gun-shot, some hint may be got from our experience in the Crimea. Out of twenty-three cases of amputation which took place, either in our army or in that of the French, not one recovered; and nearly all died miserably, very shortly after operation. All those, on the other hand, on whom excision was practised, living in comparative comfort, all without pain, for a considerable time. Out of six operated on one survived for more than a month, one died from causes unconnected with the operation, and one case recovered entirely. The *chance of saving life* is thus manifestly on the side of excision, and this is truly the most important aspect of the question. The objection so often advanced to the operation, that the limb resulting from excision is useless, even if true, has nothing to do with the matter. It is a question of deeper and more serious bearing than such an objection would imply. The only point worthy of discussion is, which operation holds out the best chance of preserving life? The little light derived from our Crimean experience is quite conclusive, so far as it goes. In the one case a life was saved, while, out of four times as many cases of the other operation, not one survived. It is true that many cases submitted to amputation may have undergone more extensive injury than any of those excised, and 'it is also true that one case of exarticulation did, to all intents and purposes, recover; yet the shock of excision must be much less than that of amputation, seeing that the great vessels and nerves are not touched, and that those changes in the blood of the limb are not interrupted, which some authorities contend is the cause of death after amputation. In all the cases of excision the loss of blood was trifling—a matter of much moment with patients like ours—and the immediate relief from pain and irritation was very marked in all the cases.

Gun-shot wounds of the hip-joint, are in many instances particularly adapted for resection, the injury of the bone being often limited, and the soft parts but little destroyed. There are, on the other hand, few accidents which present these conditions in civil life. When the shaft of the femur is split below the trochanter major, excision is hardly applicable, although Seutin performed it when he had six inches of the shaft to remove.

Seutin, Oppenheim, and Schwartz have all excised the hip for gun-shot injuries, but not with success, although all seem to have been impressed with its feasibility. Paillard gives an account of Seutin's case, from which it would appear that the patient sank on the ninth day from gangrene. Six inches of the bone were removed in this case. In Oppenheim's case, the bone was removed as low down as the little trochanter, and the patient lived eighteen days. Esmarch relates Schwartz's case. It was a secondary operation. The bone, "to two inches below the small trochanter," was removed. He died of pyœmia on the seventh day after operation. Esmarch refers to another case, operated on by Dr. Ross, and related in the 41st number of the *Deutsches Klinik*, 1850, which ended similarly. This last operation was performed two years after injury.

For disease excision of the head of the femur has been now often performed, and many times with success; the very great difference, however, that exists between the operation as performed for disease and for accident, prevents any comparison being made between their results.

Much might have been done, if we had had another campaign, to determine the exact value of excision, as applied to gun-shot wounds of the hip-joint. If the cases were selected with care, and the operation early performed, before the vital powers began to flag, and if the after treatment were carefully conducted, much might be expected from this operation in military practice. It is often very difficut to tell how far the destruction of the bone extends, either upwards or downwards; but if the case should turn out to be too complicated for excision, then amputation may be performed. Stromeyer has shown that, although the splitting of the bone barely extends into the capsule—as it did in my case— yet excision should be at once performed, as suppuration is sure to be set up in the articulation, and death by exhaustion follow. The same surgeon has also shown how it happens that, although the neck of the bone be fractured by a ball, yet the usual signs of such an injury—the shortening and rotation of the foot—may be absent, from "the fragments hanging together better on account of the partial preservation of their fibrous covering," and, in one case which he examined, a considerable power of flexion and extension remained although the neck of the bone was fractured; while, in another case, "the fragments fitted so well together that the patient did not experience the least pain, and the leg could be moved without causing crepitation." The existence of the fracture was only determined in this case by the presence of a profuse discharge. The patient himself may even be able to move his foot, and so mask the diagnosis. Esmarch gives "the extensive swelling occurring rapidly, and the pain on motion," as the only two signs which are nearly always present.

It need hardly be added, that if, in fracturing the neck of the bone, the ball or any of the osseous fragments injure the great blood-vessels, the cause is not one for excision.

The knee was only once excised, so far as I know, during the war. The operation was performed in the general hospital by Mr. Lakin, whose notes of the case have been kindly furnished to me.

"Henry Gribben, aged 19, a private in the 77th regiment, was admitted into the general hospital on September 8th, 1855. While retreating from the Redan on that day, he received a musket ball in his left popliteal space, causing him much difficulty in walking; nevertheless, he succeeded in regaining the advanced trenches, distant about 100 yards. He was a man of average muscular development, and of habitually good health. On admission, a circular wound, with inverted edges, was found at the inner part of the popliteal space, and at the level of the junction of the upper and middle thirds of that space. It was of a diameter just sufficient to admit the index finger, which could be passed to its full length in a direction forward, and slightly upward, between the inner hamstring tendons. No fracture nor other injury of the bone was detected, neither could the further course of the ball be ascertained by means of a probe or elastic catheter. It was not considered prudent to use much force with these instruments, in consequence of the close proximity to the joint, and of the absence of any satisfactory evidence that its cavity was already opened. There was no aperture of exit, the limb was not altered in shape; flexion and extension, especially the former, were limited, and any attempt to move the limb beyond these limits was attended with much pain, which was otherwise slight. Simply bearing the weight of the body only caused some uneasiness, and there was no tenderness on pressure from without. There was no appearance of synovia about the wound, nor was there any bleeding. Under these circumstances it was considered that any operative measures for the purpose of removing the ball were not justifiable."

The limb was placed upon its outer side, with the knee semiflexed, that being found the most agreeable posture, and cold dressing was applied. The patient remained almost free from pain, except when the limb was moved, and in good health, until September 20th, twelve days after the injury, when the joint became somewhat inflamed, as indicated by increased pain and heat, slight tenderness to pressure, and moderate swelling. Twelve leeches and hot fomentations were applied, and afforded great relief. The symptoms subsided, and remained in abeyance till about the 29th, when they gradually increased, the joint becoming much swollen and tender, the veins more distinctly visible, and the general health beginning to suffer for the first time, as evinced by slight perspirations, debility, frequent pulse, loss of appetite, thirst, disturbed rest, &c. The swelling of the joint was uniform, and no fluctuation could be perceived in it, though it was thought that there was some deep-seated fluctuation about three inches above the joint, on the outer side of the thigh.

"It was decided in consultation to examine the limb while the patient was under the influence of chloroform, and then to adopt such measures as the examination might indicate; accordingly, on October 1st, he was placed upon the operating table, and chloroform administered. With some difficulty, and by using considerable force, the finger could feel a part of the head of the tibia, bare and rough, a small piece of bone having been chipped off its inner and posterior edge, but the site of the ball was not detected, though it was thought to be in the joint, possibly in the space between the condyles of the femur. It was then decided to make such an incision as would admit of the performance of either excision or amputation, whichever proceeding the condition of the parts might indicate. This was accordingly done; and on opening the joint several portions of the cartilage covering both bones were found to be partially detached from the bone, softened, and their surface eroded. No fracture was found, except the small piece chipped off the inner and posterior edge of the head of the tibia."

"Excision being now decided on, and as the necessary steps were being taken, pus escaped from a cavity which existed in the outer side of the thigh, and partially surrounding the femur. The ball was found to have penetrated the inner condyle. About an inch and three quarters of the femur was removed, as well as a thin slice of the head of the tibia. The patella was also dissected out, because portions of its cartilage were softened, and partially detached. The slight oozing of blood was soon stopped by cold water." No vessel required a ligature. The edges of the wound were brought together, and retained by sutures and strapping. The extremities of the incision were left open, to allow of the escape of pus, &c. Wet lint was applied, and the limb placed in a straight position on a M'Intyre's splint, with a short whalebone splint on each side of the joint, secured by strap and buckles. The patient was placed in bed, and a grain of morphia given him.

"The portion of femur removed was about one and three-quarter inches long, and presented an ordinary round musket bullet, about half embedded in the inner condyle, the bone not being split, but the joint opened."

No symptom arose calling for remark up to the 25th, when Mr. Lakin's report runs thus:—"Had continued slowly improving and gaining strength until to-day; the discharge had diminished in quantity. Had not accumulated nor bagged. The limb had acquired slight firmness. The wound looked healthy, and had nearly healed across the front. Some difficulty had been found in keeping it in very accurate position, as he twisted about when using the bed-pan, and he is naturally a reckless, troublesome fellow. His bowels were occasionally slightly relaxed, but this was soon relieved by a dose of the aromatic mixture. To-day he seems progressing favourably, but has got his limb into a bad position; bent so as to form an angle externally. A slight discoloured patch, as of a commencing slough, on the outer side of the limb, corresponding to the position of the displaced end of the femur, at the upper extremity of the wound. The plane is readjusted, and the limb secured to it by bandages. The discharge is again rather increased in quantity. A bad sore had formed upon the sacrum, but is improving under treatment."

Again, on the 27th, the report says, "Complains rather of chilliness this morning, but has had no rigors. Has vomited several times, and his bowels have been purged. Pulse 110. Tongue moist and clean, wound healthy, small slough on outer side not extending, discharge as usual, urine drawn by a catheter."

The diarrhœa, although temporarily checked by treatment, went on, and the sickness greatly prostrated his strength. Mr. Lakin notes as follows on the 28th: "Rapidly getting worse. Pulse 130. Very low; evidently sinking; countenance much altered, but simply looking sunken and pale, and not having the peculiar aspect of pyœmia. Died at night."

Post mortem 14 hours after death. "Before removing the body to the 'dead tent,' the orderlies had taken off the splint, and the limb had been allowed to hang down, so as to destroy any points of union that there might have been. The wound had healed, except its extremities, the granulations on which had shrunk and assumed a black appearance (*post mortem*); the opposed surfaces of the bones presented a very similar appearance, and there was no sign of dead bone. They had become moulded to one another in shape. Whether there had been any union towards the centre was not evident; at the circumference there were appearances of some adhesions having been broken. The cavity of the joint contained only a small quantity of pus. The abscess in the outer part of the thigh had almost healed. No purulent deposits could be found in any of the organs, nor could any appearance of phlebitis be detected. The viscera were healthy."

"I ascertained," adds Mr. Lakin, "after his death, that, on the 26th and 27th, he had eaten some apples which he had bought, and that the vomiting and diarrhœa came on after that. He had not at all the appearance of a man suffering from pyœmia, but seemed simply to die exhausted by sickness and diarrhœa."

"The opening through which the bullet entered remained patent all the time, and a great deal of the discharge escaped through it; though probably the two extremities of the incision would have been sufficiently on the posterior part of the limb to prevent the matter from bagging."

Admiring as I do the brave attempts which have been made in civil practice to save limbs by excising the knee, I regret that it should not also be extended to military practice; but except in rare circumstances, I fear that cannot be accomplished, from the careful after-treatment and the long period of convalescence necessary to effect a cure. Ferguson speaks of more than 100 cases having been now operated on in civil practice, and Butcher has shown that the mortality is greatly less than what succeeds amputation of the thigh; but it is to be remembered that these cases were of an age and a history which rendered the procedure much more hopeful than it almost ever can be in warfare. A diseased joint is a constant source of irritation and depression to the constitution, so that, in the words of Sir Philip Crampton, "by its total excision all those parts which were diseased, and influenced the constitution so unfavourably, are removed from the system, and the injury is resolved into a case of clean incised wound, with a divided, but not fractured or diseased bone at the bottom of it," and thus the powers of the system which went to feed the disease are already so diverted to the part as to build up the loss, so soon as they can work on a proper material. That nice adaptation, however, of the surfaces, that accurate fixture of the limb, the careful attention, nourishment, and perfect repose which such cases obtain in a civil hospital, and which have so much to do with the result, can hardly be attained in the field. Mr. Ferguson, in the last edition of his admirable manual, thus sums up the advantages which his large experience ascribes to the operation: "The wound is less than in amputation of the thigh, the bleeding seldom requires more than one or two ligatures, the loss of substance is less, and probably on that account there is less shock to the system; the chances of secondary hæmorrhage are scarcely worth notice, as the main artery is left untouched; there is, in short, nothing in the after-consequences more likely to endanger the patient's safety than after amputation, whilst the prospect of retaining a useful and substantial limb, should encourage both patient and surgeon to this practice."

If the operation be performed in the field, the sooner it is undertaken the better; for, although primarily free of disease, the articulation soon becomes affected, if it be left a prey to inflammation and abscess; the constitution rapidly sympathizes, and that blood-poisoning which is so liable to follow may be established before we well see the danger of delay. Secondary operations too, it should always be remembered, do not hold out the same prospect of success in military as they do in civil practice.

The saving of blood, and the absence of any fear of secondary hæmorrhage which has been pointed out by Butcher and Ferguson, are points of much weight in favour of resection when patients are to be dealt with who are so sensitive to any hæmorrhages as those we had to deal with in the Crimea.

The resection of parts of the shafts of the long bones was not, to my knowledge, much practised in the Crimea. The lengthened period those cases take to recover, and the trying nature of this ordeal on the vital powers, made such abstinence with us almost a necessity. Several cases resulted very favourably, in which parts of the shafts of the humerus, of the bones of the fore-arm and of the leg, were thus dealt with; but in more than one case in which I knew such steps taken, too much was expected of the reparative powers of our patients, too large an extent of the bone was removed, and thus the operation failed. It was towards the end of the war that the best results were obtained from these resections. In the case of the tibia especially, the choice between amputation and resection must be guided chiefly by a consideration of the state of health of the patient, whether or not he is in a condition to withstand a long and tedious cure; by the extent of destruction of the bone, and especially of its periosteum, and finally, the means at hand for carrying out the after-treatment.

Resections in the continuity of the femur were, so far as I know, invariably fatal. The difficulty of the operation on muscular limbs must of itself predispose to disagreeable results. False joints are, as is well known, apt to occur after resection in the continuity of bones of the leg and fore-arm, when the operation is practised on only one of their two bones.

CHAPTER IX.

AMPUTATIONS.

The relative advantage of primary and secondary amputation has always held the first place among the various problems which the army surgeon has had to solve. With all that has been written on the subject by military and civil surgeons, there still seems considerable reluctance to accept the question as settled. The discrepancy of evidence brought to bear on the subject has chiefly arisen from the evident distinction being overlooked between operations undertaken for accident and for disease. Civil hospitals can seldom afford testimony similar to that obtained from the field of battle, and thus it happens that civil surgeons have come to stand in some measure in apparent antagonism to their military brethren on the point of practice under consideration. Hunter was so much of the civilian as to adhere to the consecutive operation, although, with very few exceptions, surgeons who have practised in armies have strongly advocated early interference since the days of Duchesne and Wiseman. The difference which so manifestly exists between the moral condition of the patients who are operated on for accident in civil life, and the soldier in the field, together with the circumstances in which each is treated after operation, introduce so many different items into the calculation of the question of amputation, that it is almost impossible to make use of the experience of either sphere to illustrate or influence that of the other. Besides this, the severity of those injuries which present themselves in military practice, and which authorize the removal of the limb, is so great, that it is but reasonable to suppose that an operation which removes so vast a source of irritation and pain at the earliest moment possible, must promise the best results in saving the life. In short, military experience on this point must regulate military practice, and the results of civil experience must continue to regulate civil practice.

To military surgeons, the question of primary or secondary amputation is a settled one. The experience of every war has more and more confirmed the advantages of early operation, and that in the Crimea has not disturbed the rule; in fact, later observation would lead us to go further, and in place of merely advocating interference within twenty-four hours, the prevailing idea at present would be better expressed by saying that every hour " the humane operation " is delayed, diminishes the chances of a favorable issue.

It is impossible to prove from any returns the full bearing of this question, as the mere number who survived after a given number of operations performed primarily or secondarily, by no means expresses the terms of the question. It would manifestly be necessary to know how many died before the secondary period came round, and to these should be added the victims of delayed interference, with all the pain and suffering which such delay occasioned, before we can arrive at a just estimate of the results of either proceeding. The experience in the Crimea in favor of early operation was unequivocal in both armies, and needs no illustration from me.*

Chloroform has done much to render the success of primary amputation, as contrasted with secondary, yet more marked. If we believe, as I certainly do, that by the use of this anæsthetic all fear of intensifying the shock is obviated—which was one reason why surgeons delayed operation—then the tendency of military surgery, since the introduction of chloroform, must be to still earlier and more prompt interference.

Secondary amputations were much more common during the early than the late period of the war—a circumstance which arose from the deficient means of treating the wounded in the camp during the former as compared with the latter period, and thus the necessity that existed of despatching them from camp immediately after being injured; and this, together with the better hygienic condition of the patients towards the end of the war, accounts for a fact—well known to those who served in the East, but which the range of the returns does not enable me to show in figures—that amputations were much more successful as a whole, towards the conclusion, than at the outset of the war. At first, too, when patients were early sent from camp, not a few operations, to my own knowledge, were performed during the "intermediary" period, and, without one exception, those thus falling within my observation were fatal.

The tremendous destruction which was at times occasioned by round shot or shell, left little hope from any operation whatever. In the case of many, a "pansement de consolation" was the only alternative, while, in not a few, the injury was so severe that, although amputation was performed, in the vain hope of a possible success, yet the apparent advantage of primary operation thereby suffered, and this circumstance is another of the many which makes it impossible to place this question in a fair light. The most severely injured have their limbs removed early, while the most hopeful are retained for secondary operation, and thus all the advantages of slighter injury—less consti-

---

* I am led to understand, from a very well-informed source, that the Russians also lost two-thirds of all their secondary operations, but saved a fair number of their primary.

tutional disturbance, more promising habit of body, and state of general health—are denied to the early operations. In truth it may be said, that if, with all the advantages under which secondary amputations are recorded, they appeared as merely equal in success to primary, then the superior claims of the latter to our attention would be sufficiently clear; how much more marked, then, are the successes of early operations when we find them giving such superior results!

As to the general success of amputations during the late war, it may be safely said that when due weight is given to the many circumstances which have militated against the success of all operations, and which have been fully dwelt upon in the course of the preceding pages, those performed early have afforded a very fair proportion of success; while it cannot be denied but that those undertaken late have been followed by most unfortunate results.

A siege presents peculiarly favorable opportunities for testing the value of *immediate* amputations. The men being close together, and acting within a narrow space, can be seen almost instantly on being injured. The position of the soldier in such circumstances resembles that of the sailor on board of his ship; so that the experience of naval surgeons, which is so strongly in favor of instant amputation, applies with peculiar force to military siege practice. Unfortunately, the arrangement followed in our army during the siege of Sebastopol, made the elucidation of this point impossible. Assistant surgeons were alone sent to the trenches, (except during an assault, when a staff surgeon occupied one of the ravines behind each division; but in the hurry and confusion which prevailed at such times, the men he operated on were lost sight of;) and as by the rules which prevail in our service an officer of that rank is not allowed to amputate, except when the surgeon is not with the regiment, no means existed for the due examination of this question. The French experience, if it were available, would be of much use on this point, as they performed many capital operations in their trench ambulances.

Whatever that condition is which is conventionally known as "shock," it seems pretty evident, from the admission of all, that it is not established for some little time after the receipt of injury—an interval which differs in duration, mainly in accordance with the severity of the wound, the agency by which the injury has been caused, and probably the constitution of the sufferer. The evidence of naval surgeons, as summed up by Mr. Hutcheson, in reference to the absence of shock immediately after the receipt of a wound, must be conclusive to all unprejudiced minds; and instances were not wanting during the late war which appeared to support the same view. I know of several well-authenticated cases which occurred during the siege, in which the perfect absence of all constitutional prostration after an accident so severe as the carrying off a limb, and the non appearance of such shock for some considerable time after, went to prove the same position. If this precious moment could be seized at all times, and that operation performed under chloroform, which assists so much in warding off the "enbranlement" we fear, how much more successful would our results prove, than under any other circumstances they ever can be!

It is during this interval, too, that we obtain the full good of the soldier's *moral* advantages over the civilian. "Cut off the limb quickly," says Wiseman, "while the soldier is heated and in mettle"—and the observation is as old as Pare, that while excited by the combat, and yet within sound of the cannon, the soldier or sailor willingly parts with a limb which a few hours of reflection would make him desire to run the risk of preserving, and upon which he fixes all his attention, so as to magnify greatly the dangers of the subsequent operation. Moreover, the removal of the man, before operation, to any distance from the scene of his accident, lessens somewhat his chances of recovery; as, besides the danger that the irritation and pain of such transport, however carefully it may be conducted, will occasion—the constitutional depression we dread ; the mere loss of blood which, although going on in very small quantities, is yet flowing in drops, when a drop may extinguish life, are serious objections to the shortest delay.

But even although that constitutional disturbance which is the result of injury is present, is it always necessary to wait its subsidence before operating? If it be very decidedly marked, and the patient thus much prostrated, such delay may certainly be called for ; but it is an opinion often stated by those who must be well informed on the subject, that such delay is not always advantageous, but manifestly the reverse. Larrey, for example, gives repeated utterance to the following sentiment : " Il est donc demontré que la commotion, loin d'être une contre-indication à l'amputation primitive, doit y determiner le chirurgeon ;" and again, " Les effets de la commotion loin de s'aggraver, diminuent et disparaissent insensiblement apres l'operation ; and in this opinion he is by no means solitary, as may be seen by reference to the writings of many naval surgeons, who have manifestly the best opportunities of judging in the matter. The upholding influence of chloroform comes strongly into play in such places, and obviates, in a great measure, the dangers which have been prognosticated from such proceedings. If the constitutional depression be the result of an injury which remains as a source of irritation, then the removal of such must manifestly be a great point gained; and I know it is the opinion of many army surgeons of large experience, that the presence of shock is no hinderance to operation (under chloroform) if that condition be not very decidedly marked at the moment of interference.

The difficulty which chiefly stands in the way of instant operation is the recognition of the cases which demand it, and the certainty that no fatal internal lesion may not have been at the same time sustained, as the accident to the limb which necessitated its removal. However, it would certainly tend on the whole to the saving of life to operate as soon as possible, not only in all those cases in which the necessity for it was evident, but also in all doubtful cases; as although a few limbs might thus be be sacrified, I have not the least doubt but that many lives would be saved.

The Crimean was afforded a most excellent field for observing the relative value of flap and circular amputations; as, although in our army the former was commonly employed, most of the French and not a few of our own surgeons adhered to what Sir C. Bell termed "the perfection of the operation of amputation." As the advantage, in general, of removing the limb as far as possible from the trunk is fully recognized, it seems curious that the circular mode of operating, which I think admits of this more than the operation by flaps, should not be more followed. In the lower part of the thigh this is particularly observed. Protrusion of bone is the great bugbear which terrifies most operators: hence they make unnecessarily long flaps, and remove a much larger amount of the bone than is at all necessary. This was very apparent in many amputations in the East. Mr. Syme has laid down the true principle which should regulate our proceedings, when he says, "It is not the length of the flaps which prevents the risk of protrusion of the bone, but the height at which it is divided above the angle of union of the flaps."

In soldiers, as in many (although not all) cases submitted to primary amputation for accident at home, the proportion of muscle to skin and subcutaneous fat is different from what it is in most cases operated on in civil hospitals, and thus modifies our appreciation to some extent of the two modes of operating. In soldiers there is commonly but little subcutaneous fat, and the muscles are large and strong; hence it becomes very difficult, when practising the flap operation, to adapt the parts to one another, so as to fulfil the latter part of the old maxim, "Muscle must cover bone, and integument muscle." It cannot be said that this arose in the East from the maladroit performance of the operation by the flap, as the same circumstance may be seen to occur at home in the hands of our ablest hospital surgeons. The paring and stuffing-in processes which are not uncommonly seen in hospitals, to correct the results of the condition referred to, are no less prejudicial than unsightly. The irritation is thereby increased, and proper adhesion of the parts prevented. In secondary amputation the excess of skin removes any fear of similar accidents. Chloroform has refuted the argument in favour of the flap operation, founded on the greater speed of its permance than the circular, as such great speed is now a matter of no moment. But however it be with regard to the question in general, there is one fact which any one who had opportunities of watching matters during the early part of the late war will amply verify, viz., that the circular stumps stood the transit to the rear much better than those formed by the flap method, and thus it would seem that the former mode of operating is more advantageous in military practice than the latter. The long heavy flaps were so knocked about during the land and sea passage, that they often became loose, got bruised, and ended by sloughing; while the firm, compact stumps made by the circular method were little, if at all, injured. When patients can be treated in camp to a termination, the influence of this circumstance is, of course, null. It may be said that the length of the flaps was a mistake committed in the operation; but, unfortunately, such errors must always be looked for in like circumstances, where there is a large body of operators, most of them without previous experience in operating, and whose chief fear always is to have "too little flap;" for although it is true what Hammick says, that "it requires more practical experience to know when to take off a limb than how to do it," yet the *how* must also be studied, like everything else.

In considering the statistics of amputation performed during the Crimean war, the figures refer solely to the period between the 1st of April, 1855, and the end of the war, and consequently exclude all the unfavourable part of the campaign, as well as the greater number of the operations which were absolutely performed during the war. It was found impossible to attain to accuracy with regard to the earlier period, so the field of observation was restricted as stated. It is needless to point out how different must be the lessons derivable from the statistics of this latter period alone, to what they would have been if the whole period of the war had been included.*

During the limited period I have mentioned, there were 732 amputations in all parts performed, followed by death in 201 instances; of these, 654 operations and 165 deaths were primary, and 78 operations with 36 deaths, secondary; giving a per-centage of 27·4 deaths overhead—25·22 for the primary, and 46·1 for the secondary operations. If we include only the greater operations, viz., amputations of the shoulder, arm and forearm, of the hip, thigh, knee, and leg, then we have a total of 500 cases and 199 deaths, or 39·8 per cent.; of which total 440 cases and 163 deaths, or 37

---

*In my original papers the figures were intended to represent the period of the whole war. I have reason to think, that although, upon a more careful investigation of the returns than could be made in the Crimea, these numbers have since proved not strictly accurate, they yet represent pretty much the results which followed many of the operations as viewed in the more lengthened and less favourable aspect of the war.

per cent. were primary, and 60 cases and 36 deaths, or 60 per cent. were secondary.

The increase of the mortality as we approach the trunk may be shown thus, taking the primary amputations alone as giving the most unbroken series:—

SUPERIOR EXTREMITY.

Part.	Ratio mortality per cent.
Fingers,	0·5
Forearm and wrist,	1·8
Arm,	22·9
Shoulder joint,	27·2

INFERIOR EXTREMITY.

Part.	Ratio mortality per cent.
Tarsus,	14·2
Ankle joint,	22·2
Leg,	30·3
Knee joint,	50·0
Thigh, lower third,	50·0
" middle,	55·3
" upper,	86·8
Hip joint,	100·0

The lower extremity was removed at the hip joint seven times during the period included in the returns, and at least three times more previously, giving ten cases, all primary operations, and all ending rapidly in death. One of these cases was operated on by my lamented friend Dr. Richard M'Kenzie, after the Alma. The French had thirteen cases, primary and secondary, after the Alma and Inkerman, and all died. One of these, a Russian, was operated on by M. Legouest on the 3rd of October, 1855, at Constantinople. The upper part of femur was completely smashed by a conical ball. The flaps had adhered to a point by the middle of December, at which date I saw the patient walking about the ward on crutches, and looked upon by all as being beyond danger. The very night on which the order arrived for sending him to France—where he was to be admitted, by special permission, into the Val de Grace—he fell when walking in the corridor, and hurt his stump so that it bled profusely. Inflammation was set up, suppuration, renewed hæmorrhage, and diarrhœa followed, and he died on the 9th of February, four months after operation. M. Mounier in the same hospital had three cases, one of which I watched with interest. Two of these died of hæmorrhage, one on the fifteenth, and the other on the twentieth day. The third died of cholera. One of these men was a Russian.

The mortality which has thus followed exarticulation at the hip during the Eastern campaign, has been very deplorable; yet, in the cases in which it was performed, no other alternative remained, except to abandon them to inevitable death, which many might be disposed to think the more humane proceeding, as they often linger for a long period before death. M. Legouest's case was unquestionably successful; and, although we can hardly hope with Larrey that this operation will ever be performed as readily as his favourite one at the shoulder joint, still the results of operation at the hip for accident have not been so utterly hopeless as to lead us to abandon it. M. Legouest has given, in a most interesting paper on the case mentioned above, a table containing most of the recorded cases of amputation at the hip for gun-shot wounds. Of primary operations he has collected 30 cases, all ending fatally; of intermediate or early secondary operations he finds mention of 11 cases, with 3 recoveries; and of operations performed at a period so late as that "the injury had lost all its traumatic character," 3 cases, with one recovery. Thus, if we sum up the whole, we have 4 recoveries in 44 cases, or a mortality of 90·9 per cent. Some of the primary cases died on the table; all of them before ten days except 2, which perished within a month. The proportion of recoveries among those operated on after the primary period, but before a long elapse of time, *i. e.*, at some period during the existence of "the traumatic phenomena," was the largest, and hence that would seem the best time to undertake the operation.

During the Schleswic-Holstein war, amputation at the hip was performed 7 times—5 were operated on by Langenbeck; only 1 of these cases recovered. I find no mention whether these cases were primary or secondary. In the Indian campaigns I find mention of only 1 case of amputation at the hip for a gun-shot wound. It was a primary operation, and took place in the Punjaub. Thus, if we reckon the whole number of cases operated on for gun-shot wounds, those recorded by Legouest, our own Crimean cases, and the Holstein and Indian ones, we find a total of 62 cases, and 5 recoveries, or a mortality of 91·9 per cent.

Mr. Sands Cox, recording the experience of civil hospitals, as well as those of military practice, up to 1846, gives in all 84 cases, most of them for injury, with 26 recoveries; 14 of these successful cases being after accident, and of the unsuccessful, 20 were for injury; and in the *Medical Times and Gazette* for April, 1857, there is a further record of 8 cases, of which 2 were for accidents, (1 primary and 1 secondary,) with three recoveries, all after operations for disease. Cox recognizes the difficulty of restraining the hæmorrhage during the operation, and the shock given to the nervous system, as the great sources of danger. The hæmorrhage, at a considerable period after operation, would appear even a more common cause of the fatal event, than the difficulty of commanding it at the time.

It will, of course, only be in the event of such destruction to the bone or soft parts, or such other injury to the nutrition of the extremity, as puts resection out of our power, that amputation will be performed. If the fracture of the neck of the bone were slight, as when occasioned by a small ball, or one striking with little propulsive force, such as that projected by the matchlock, then the case, I conceive, must be viewed more as a compound fracture of the upper part of the thigh, and should be treated accordingly. M. Legouest has recorded 6 cases in which the limb was not removed or resected, and 3 of these recovered. One of these cases of recovery having occurred in 1812, must have been wounded by a round ball; the second was injured in a duel, and hence probably by a small light ball; while the third was observed in Africa, where neither the size nor the form of the balls used by the natives is to be compared to the conical bullet. All three were struck on the trochanter. The 3 fatal cases with us which were not interfered with, took place after the Alma and Inkerman, and hence were probably wounded by conical balls.

All are agreed that, when practical, the separation of the limb should be accomplished at or through the trochanter, rather than at the joint, on account of the diminished risk ; and this can be more often executed than would at first appear, as it not uncommonly happens that the fracture does not extend to the head of the bone, as it seemed at first sight to do ; hence it might be judicious, in all doubtful cases, to make the incisions so low as to suit amputation at the trochanter. The steps necessary for exarticulation can easily be taken, if called for afterwards, when the bone is examined. Such a proceeding would certainly not be very " brilliant," but it might save a life.

After the 1st April, 1855, amputation in the upper third of the thigh was performed 39 times, with a fatal result in 34 cases. Of the total number only one was a secondary operation,- and it ended fatally. The ratio mortality per cent. was thus 86·8 for primary, and 100· for secondary. I have never myself seen any case recover in which the limb was amputated *beyond doubt* in the upper third, and I never met any one who had, except in one instance, and that man was seen in England. I saw several upper third amputations, so-called, which was not really so. It is very easy to be deceived on this point. The French and Russians found these operations so hopeless that they almost abandoned them ; and in fact, as was before remarked, the attempt to save such limbs, hopeless as it was, seemed more promising than amputation in the field.

Amputation in the middle third was performed during the period after the 1st April, 1855, 65 times, of which number 38 died ; 56 of these cases and 31 deaths were primary operations, giving a ratio mortality per cent. of 55·3 ; 9 cases were operated on at a late period, and 7 died, or 77·7 per cent. Amputation in the lower third was performed during the same period 60 times, 46 being primary, and 14 secondary operations; of the primary, 23 or 50· per cent. died; and of the secondary, 10 or 71·4 per cent. A very great many of the operations classed as "lower third," ought to have been entered as "middle third," as it very frequently happened that, from the operator adhering too closely to the maxim of Petit, to "cut as little of the muscle and as much of the bone as possible," an operation which was ostensibly in the lower, was in reality in the middle third. This is a matter of which I have seen many illustrations; consequently, I believe that at least one-third of the operations and the deaths classed as lower third, should be transferred to the middle third column, and thus the relative frequency and fatality of the two operations would be better expressed.

Taking amputations in all parts of the thigh, then, we find the number of operations after the 1st of April, 1856, was 164, of which number 140 were primary, and 24 secondary operations. The total mortality was 105, or 64· per cent. Of the total deaths, the primary amputations yielded 87, or 62·2 per cent, and the secondary 18, or 75· per cent. It must always be borne in mind, that these results only refer to the period of the war when, as was before stated, secondary operations were becoming very rare, and the state of matters in camp so improved, that the total mortality after amputations was by no means what it had been at an earlier period; so that to say that the average mortality after amputation of the thigh in the Crimea was 64· per cent., does not by any means express the whole truth. However, if we take the later period only into consideration, then our results may be thus contrasted with those obtained in other fields of observation.

AMPUTATIONS.

TABLE SHOWING THE PER CENTAGE OF DEATHS AFTER AMPUTATION (PRIMARY AND SECONDARY) OF THE THIGH FOR GUN-SHOT WOUNDS AND ACCIDENTS.

	Mortality per cent.
Crimea, British army from April 1st to end of war,	64·0
Constantinople, French Dolma-Batchi hospital, Mounier,	82·6
Naval Brigade, Crimea,	65·0
Indian campaigns,	48·7
Waterloo,	70·2
Spain, Alcock,	62·0
Schleswic-Holstein, Esmarch,	60·15
Danish army, 1848–50, Djorup,	56·7
Sedillot, "Campagne Constantine," 1837,	87 5
Africa, Baudens,	51·4
Polish campaign, Malgaigne,	100·0
Mexican War,	100·0
Hotel Dieu, 1830,	81·8
Cases communicated to the Academy, 1848,	77·2
INJURY.	
Phillips,	71·8
Parisian hospitals, Malgaigne,	73·9
Glasgow, previous to 1848, Lawrie,	75·0
" M'Ghie,	78·6
" Steele,	72·0
St. Thomas' hospital, South,	85·7
Hussey,	62·5
James, *all primary*,	61·5
University college, Erichsen,	60·8

The usual discrepancy which marks statistical tables is observable in the above enumeration. That between the results obtained in our army, and those quoted from the French, and which were kindly furnished to me by M. Mounier, is easily understood, when it is stated that of the total number of 46 amputations of the thigh which presented themselves in the hospital presided over by that distinguished surgeon, 25 were secondary operations, all of whom perished, while in our returns, and those of the Naval Brigade, there were very few consecutive amputations. Out of 21 primary amputations reported by M. Mounier, 8 recovered. The low mortality among the Indian cases is somewhat difficult to account for. In calculating them, I did not include any case except those the result of which I could find well authenticated. To distinguish between primary and secondary operations, in many of the cases recorded by the various authors referred to in the above table, was found impossible, but so far as this can be accomplished appears in the following table:

TABLE SHOWING THE MORTALITY AFTER PRIMARY AND SECONDARY (DISTINGUISHED) AMPUTATIONS OF THE THIGH FOR GUN-SHOT WOUNDS.

	Mortality per cent.	
	Primary.	Secondary.
Crimea after April 1, 1855,	62·	75·
Constantinople, Mounier,	61·9	100·
Legouest,	··	100·
Naval Brigade,	66·	60·
Indian campaigns,	38·	69·
Spain, Alcock,	64·7	60·
Africa, Baudens,	13·3	80·
Cases communicated to the Academy, 1848, in which the distinction is drawn,	57·	81·2

TABLE SHOWING THE MORTALITY AFTER PRIMARY AND SECONDARY (DISTINGUISHED) AMPUTATIONS OF THE THIGH FOR INJURY.

	Mortality per cent.	
	Primary.	Secondary
Malgaigne	75·	60·
Glasgow, Lawrie,	91·6	66·
" Steele,	65·6	83·6
" M'Ghie,	61·2	96·6
St. Thomas' hospital, South,	100·	50·
University college, Erichsen,	57·	62·5
Hussey,	83·	··
James,	61·5	··

If a calculation is made of the mortality succeeding amputation of the thigh from gun-shot wounds *alone*, and the whole number of cases referred to in the above table included, then the average mortality per cent. of primary operations would appear to be 56.5, and of secondary 79.0, while, if the operations performed in civil hospitals for injury are alone calculated, then the average mortality of primary operations would appear as 69.6 per cent., and secondary 75.4, a result somewhat different from what is usually obtained.

Amputation through the knee-joint has been performed in our army 6 times primarily, 3 of which were fatal, and once secondarily with a fatal result. This very old operation has lately been creating some interest in the profession, and was often performed by the French surgeons in the Crimea. The opinion they were led to form of it may be supposed to be expressed by Baudens, when he says (*Une Mission Medicale en Crimée*), "It is a truth which the numerous facts observed in the Crimea permit us to affirm, that, whenever it is impossible to amputate the leg, the disarticulation of the knee should be preferred to amputation of the thigh. The former has more often succeeded than the latter." There are not, however, very many cases occurring in the field which are adapted for this operation, as it should be performed only when the injury is limited to the leg-bone, and the femur remains intact; and when this takes place, it often happens that the soft parts are so much implicated as to deprive us of flaps. However, if the posterior flap is destroyed, we can take a long flap from the front, and *vice versa*. To 4 of the cases operated on in camp, with the details of which I am acquainted, the operation was not applicable, as the femur was more or less injured, so as to call for the removal of part of it: hence the operation, although termed amputation through the knee, was in reality low amputation of the thigh, such as that now employed in white swelling of the articulation.

As to the mode of operation, the French mostly adopted Baudens' method, but in 5 cases operated on in the general hospital that proceeding was departed from, in so far as that the posterior flap was made from within outwards in place of the reverse, as directed by that well-known surgeon. The anterior flap, too, was not made so long. Whatever method of operating be adopted, the great point which demands attention is, to have the flap sufficiently broad to cover the expanded end of the femur, which there requires a large and broad covering. Of the 5 cases operated on in the general hospital, one died of phagedænic sloughing on the forty-third day; another, a soldier of the 62nd, died of enteritis on the sixty-seventh day, the stump being healed to a point; a third sank from exhaustion on the ninth day after operation; a fourth never fairly recovered from the shock; while the fifth and last case recovered, under the charge of Dr. George Scott, who operated on him. This patient, a soldier in the Buffs, was struck on the right knee-joint by a ball, on the 8th of September. He thought himself very slightly injured, as the only thing he observed wrong with the joint was his inability to flex it, on account of "something catching in it." A small opening was found in the middle of the popliteal space, slightly external to the middle line, from which a good deal of blood flowed. This opening led into the cavity of the articulation, and spiculæ of bone were felt within. A part of the end of the femur was removed, but the patella left. A round ball had pierced the external condyle, and lodged. The posterior flap eventually sloughed, and exposed to the end of the femur; but the bone became subsequently covered over with granulations, and though the patient's progress towards recovery was much impeded by the formation of an abscess among the muscles of the thigh, which required extensive incisions, he went to England in perfect health in January. His stump was strong and firm, and he had much power over its movements. The patella could be felt on the upper surface, to which position it had been gradually retracted. In several of the cases which I have seen in the French hospitals, where sloughing of the flaps had taken place, as in this case, and exposed the extremity of the femur, the cartilages were alone thrown off, but not a scale of bone.

So far as I can judge, the practical advantages of this operation are equal in value to those theoretical ones which its advocates claim for it, and they would seem to recommend its more general adoption in any future campaign. First of all, the shock to the system is less, and we obtain a larger and firmer stump than when the femur is sawn through; the end of the bone on which the patient has to bear his weight is likewise more expanded, and more rounded, and hence calculated to inspire greater confidence in the patient in the use of it, and less liable to cause ulceration by its pressure on its coverings.* A false leg can be more easily attached to such a stump, and more power is retained in progression from the muscles which remain undivided, than when the limb is amputated in the continuity. Few now participate in Liston's opinion of a long thigh stump, but, on the contrary, most surgeons try to keep their section as far as possible from the trunk. The non-interference with the medullary canal obviates many of the dangers of amputation, according to Cruveilhier; while the extremity of the femur, which is largely supplied with blood-vessels, being retained, there is less fear of exfoliation than when the dense tissue of the bone

---

* The absorption of the condyles of the femur which may go on after this operation, is illustrated by a case mentioned by M. Legouest (*Amputation partielles du pied*), in which a soldier had undergone amputation at the knee in 1800, in Italy, and "the enormous tuberosities had so diminished in volume that no trace of them could be recognized, but the member presented a cone terminated by a point." So completely had the part changed, that it was only after very careful examination they believed the man's story, that he had been amputated at the joint.

has been opened by the saw. The position of the divided artery in the centre of the flaps, and the few ligatures which required, are further arguments in favour of this operation. There is little fear but that the flaps will adhere over the cartilaginous extremity of the bone—in fact the cartilages soon disappear during the healing process. There is some appearance of force in the objection which some have advanced to the operation, that from the length of the stump no proper space is left the play of an artificial joint; but if it be evident, as civil statistics at least prove, that the fatality attendant on this operation is less than that which follows amputation of the thigh, then any such objection loses all its weight.

If then, cases were selected for the operation in which the femur remaining intact, and the legbones being destroyed, a sufficiency of flap could be got from the calf, or the front of the leg, and if the amputation was performed early, I firmly believe, with Malgaigne that it is " Encore une de ces operations trop légèrement condamnées, et qui lorsqu'on a le choix mérite toute préférence sur l'amputation de la cuisse dans la continuité."

The leg was amputated after April 1, 1855, 101 times, with death following in 36 cases, giving a mortality of 35.6 per cent.; 89 cases, and 28 deaths, were primary operations, and 12 cases, with 8 deaths, secondary—thus affording a ratio of mortality per cent. of 31.4 for the primary, and 66.6 for the secondary.

The rule generally followed in our army, has, I think, been to preserve as much as possible of the limb, but except in those cases in which the operation was performed just above the ankle-joint, the French appeared usually to amputate at the place of election. I saw no instance in which Larrey's operation through the head of the tibia was had recourse to, but I am informed that it was several times successfully performed in the French ambulances.

The greatly improved mechanical contrivances of late years have much changed the bearing of the question with regard to long leg stumps. The facility and moderate cost with which artificial limbs can now be fitted to any part of the limb, from the knee to the foot, has obviated many of the reasons which formerly induced surgeons to prefer the high operation. Larrey's, through the head of the tibia, is a most valuable one when the destruction has extended high up the leg, as it enables us to retain the use of the knee-joint, as well as diminish the risk to life. That at " the place of election" will, of course, continue to be employed in cases of injury above the middle of the leg; but when the nature of the accident permits of it, the part of the leg which appears to combine most of the advantages sought in leg stumps by both the surgeon and the mechanician, is undoubtedly that in the centre of the middle third. The length of the lever thus obtained, the diminished bulk of the part and consequently of the truncated section, the means of covering the bones, and the room it affords for attaching a limb, are all in favour of this locality. Many most admirable stumps were made in this part of the limb during the war. In operations for accident, as in gun-shot wounds, we can, of course, operate lower in the leg than we can when the operation is undertaken for disease, from the absence of the thickened state of the bone, and the changed and bound down tissues which are so common in cases operated on in civil hospitals.

As to the operation just above the ankle, which has of late years caused so much discussion on the continent, we had, so far as I know, no experience in our army; but the French had a good number, which, so far as the condition of the stumps went, were by no means promising. This operation, although revived by the improved method of procedure introduced into practice by M. Lenoir, is yet of sufficiently old date. It is mentioned by Dionis in his "Cours d' Operations," and was practised by Bromfield in 1740, and afterwards by White, Alanson, and Bell, in England. In France, Blandin often performed it in recent times, but was induced to abandon it, like many others, from the bad results his method of operation yielded. By M. Lenoir's modification,* and M. Martin's artificial limb, the operation promises again to come into favour. This operation appears to me to have a special bearing on military practice. Its value will be best judged of by considering, 1st, its safety, and 2nd, the usefulness of the resulting stump. As to the first point there can be no question as to its advantage over any other amputation in the leg. The greatly diminished bulk of the soft and hard parts at the place of section, the smaller amount of shock such severance will occasion, and the more rapid closing of the wound, are all incontestable. Its fatality in the cases operated on in France has been only as one-sixth or one-seventh, while the mortality of amputation at the place of election is more than one-half (55 in 100 according to Malgaigne). In some hospitals, as in the Beaujon, the mortality has been even less in the *sus-malléolaire* operation than that mentioned above: thus M. Huguier only lost 1 out of 14 cases. So then, as far as the mortality goes, there can be no division of opinion, as there is about the second point, viz., the state of the stump afterwards. The difficulty of retaining enough of covering for the bones, the fear of such retraction as will occasion a conicity of the stump, the tenderness of the cicatrix, and its inability to stand pressure, the chance of fusiform collections of pus forming among the tendons, of caries or necrosis of the bones following,—all these are among the objections which have been advanced to the operation. If we, however, carefully examine these by the light of the large num_

---

* See Arch. Gen. de Med., July, 1840, and Mémoire by Arnal and Martin, Paris, 1842.

ber of observations which can now be brought to bear on the subject, we find that the only objections which are of any weight are the scanty covering of soft parts, the tenderness of the cicatrix, and the risk of necrosis. Purulent collections can be easily avoided by careful dressing; and the presence of the other evils, and, in fact, the want of flap also, must be referred to the manner in which the operation has been performed. I have examined a considerable number of those amputated in Paris, and am bound to say that, while in some cases the evils spoken of existed, in the greater number of instances good and firm stumps were formed. This was especially the case in several which I saw in M. Lenoir's service, in the Neckar. Some of the cases which had been operated on in the Crimea were certainly very bad. At the Society of Surgery I saw an Arab, shown by Baron Larrey, both of whose limbs had been removed above the malleoli, in the East. They were both secondary operations, and seemed to have healed well at first; but the cicatrix afterwards ulcerated, and at the period he was shown to the society (nearly two years after operation), he could not use his stumps in any way, from their being in an unhealthy condition. In another case, shown to the same society on a subsequent occasion, the operation had been performed in 1848, and the man had been an inmate of hospitals on several occasions during the interval, on account of ulceration, abscesses, and necrosis in his stump. The bones were much thickened, and evidently diseased at the time I saw him. A letter from M. Hutin of the Invalides, which was at the same time read, stated the results of the operation as they had come under his observation, and certainly his evidence was not favourable; however, the want of a properly constructed artificial limb for the patients, detracted much from the value of his remarks. If the limb cannot be fitted with a false foot, but made to rest on the knee, scarcely anything will make amends for the long and cumbrous stump. Since 1845 M. Hutin had had 5 cases especially under his notice: one could walk, but with difficulty, and would willingly part with his foot; one had been several times in hospital from the state of his stump, and three had to undergo subsequent amputation. Now all this is sufficiently distressing and discouraging, but in military practice I question whether it is conclusive. The limited mortality yet presents itself to us as a great fact, which arrests our attention. If when men die so fast after the ordinary amputation of the leg, as they did during the early part of the war in the Crimea, it becomes a grave consideration whether, with all its subsequent drawbacks, we should not adopt this process when practicable. If our choice lay between two operations of equal gravity, then unquestionably we are bound to select that which will provide the most useful stump; but when the chances of death are beyond all comparison greater in the one case than in the other,—when, independently of those dangers which attach to the operation itself, the marked presence of an "hospital epidemic" makes it desirable to expose a small and as rapidly-healing a surface as possible, then I think it may be conceded that the *sus-malléolaire* operation has many claims upon us. Life must be our chief concern; convenience a subordinate consideration. The complaints of patients about the inconvenience of their stumps, must be considered as affording little evidence in the matter, as the fact that they survive to murmur is often due to the very operation against which they complain.

If the heel remains, then this operation could not be thought of, but it is in those cases, sufficiently frequent in their occurrence, in which the whole foot has been carried away by round shot, or such like accident, and in which the choice of operation lies only between the amputation above the mallcoli or higher up, that the merits of this method can be weighed. The careful study of those cases in which caries or necrosis has appeared in the bones of the stump after the *sus-malléolaire* amputation, will be found to have been submitted to the operation for disease, and not for injury, and it will generally be found, besides, that a faulty apparatus has been used afterwards. Everything depends on the careful adaptation of the false foot, and, so far, this is of itself an objection to the operation being performed on the poor; but the view alone I wish to take of it at present, is with reference to military practice, and there it seems to promise many advantages at times when there prevails a high mortality after operations.

Amputation at the ankle-joint was performed 12 times in the Crimea during the period embraced by the returns, and death followed in 2 cases. Of the total number of cases 3 were secondary operations, and these were all successful. Syme's operation was as useful and as successful in its results as usual. Pirogoff's modification of Syme's method was, I understand, several times tried at Scutari. I saw none of these cases, and am ignorant of the results. In England it appears to have been recently followed by good effects in 6 out of 9 cases in which it was performed. Langenbeck is said to approve of its results in a good many cases in which he has tried it; but the history of the 3 cases first reported by M. Pirogoff himself, and those more recently put on record by Michaelis of Milan, and various German surgeons, does not hold out much encouragement to repeat the operation, not only from the long period necessary to a cure, but also from the unsatisfactory nature of the resulting member. It was reported in the East that this operation had been frequently performed by Pirogoff himself in Sebastopol, but that he had found the calcaneum act as a foreign body in the stump, and was hence disposed to abandon it. Roux of Toulon's operation was performed once in the general hospital in camp, with most excellent results. The chief objection to this operation arises from the vessels and nerves being drawn under the bone; however, it certainly enables us to form a stump little inferior to Syme's, when the half of the heel has been destroyed. Baudens is said to recommend the flap to be taken from the interior surface of the joint, or even from its external surface, if it can be got no other where, rather than go above the ankle. Chopart's operation was performed primarily 7 times, one case ending unfavourably, while Lisfranc's was successful in the 4 cases in which it was tried. The step now always followed by Mr. Ferguson, of removing the projection of the astragalus in performing Chopart's operation, is an undoubted improvement.

The upper extremity has been removed at the shoulder joint, between the 1st of April, 1855, and the end of the war, 39 times, with a fatal issue 13 times, or 33·3 per cent. Of these operations 33 were primary, and 9 deaths followed, giving thus a mortality of 27·2 per cent.; while of 6 secondary operations 4 died, or 66·6 per cent. During the previous period of the war at least 21 other cases of amputation at this joint were performed, beyond the 39 mentioned above, and of that number 6 died, thus presenting a total of 60 cases and 19 deaths, or a ratio of mortality of 31·6 per cent. overhead.

It is impossible fairly to contrast the results of amputation at the shoulder and that in the shaft of the humerus; as, in military practice particularly, it very much oftener happens that the trunk has suffered severely in those injuries which necessitate exarticulation, than those in which amputation of the upper arm alone is required. Not a few illustrations of this occurred in the Crimea. Thus, in at least two of the cases returned as shoulder-joint amputations, besides the injury to the arm, the scapula was carried away or destroyed, and the muscles of the chest torn.

In no operation is the advantage of primary over secondary amputation so evident as in that at the shoulder-joint; early operation at this part being an exceedingly successful undertaking, while late interference generally affords a considerable mortality. Thus, if we take Guthrie's experience in Spain, and Dr. Thomson's observation after Waterloo alone, this point is well illustrated: of 19 cases of secondary amputation mentioned by Guthrie as having been performed between June and December, 1813, 15 died, while of an equal number who were operated on in the field, only 1 died. Dr. Thomson again says, "In Belgium almost all of those recovered who had undergone primary amputation at the shoulder-joint, while fully one-half died of those on whom it became necessary to operate at a late period." The same point is illustrated to some extent by our Crimean results, less than a third of the primary, and two-thirds of the secondary perishing.

Deputy-inspector Gordon had one case of recovery, in which both the arm and the greater part of the scapula were removed. Mr. Howard of the 20th regiment successfully removed the right arm of one man and the left of another, in close succession, at the joint, for injury occasioned by the same cannon ball which had struck between them.*

Amputation of the upper arm was performed in the Crimea, from April 1st to the end of the war, 102 times, followed by death in 25 cases, the mortality per cent. being thus 24·5. Of the total number, 96, and 22 deaths were primary operations. The ratio of the mortality was thus 22·9 for the primary, and 50·0 for the secondary operations.

The fore-arm was amputated during the same period 52 times primarily, and the hand at wrist once, with only one death; while of 7 secondary operations in the same parts, 2 died.

These returns do not speak of a considerable number of secondary amputations of the arm, which were performed early in the war, and the success of which was certainly such as to warrant us in trying to save, in the first instance, most cases of gun-shot wounds of the arm. It is almost impossible to say what wound of the arm by a ball will not recover; so that it is a well-recognized rule to wait, in all but desperate cases, and only amputate if unavoidable at a subsequent period. In military practice secondary amputations are only justifiable when performed on the upper extremity.

The mode of managing stumps in the East was that usually followed at home for the promotion of adhesion by the first intention. The edges of the flaps were usually united by suture. The observation of this method in the Crimea did not certainly appear to be satisfactory. To wait, as Liston so strongly advocates, till all oozing has ceased from the cut surfaces, is unquestionably a most useful precaution, and one of great moment to their successful and early union. The irritation which the stitching of the edges occasions, the want of sufficient room for subsequent swelling, the confinement of pus which is thereby

---

* The following is a most instructive case, as showing how the operation of amputation at the shoulder may be recovered from under the most unpromising circumstances. It occurred in the 29th regiment, serving in India, and under the care of Deputy inspector Taylor. Sergeant Ritchie was struck by a cannon ball on the upper part of his left arm, by which the bone, including the head and upper third of the humerus, was smashed. Both folds of the axilla were carried away, and the artery was divided. The arm was only kept attached by a portion of the deltoid, and the skin covering it, and of these the flaps were made. This man lay exposed on the field for three days; yet he recovered completely. "His case is peculiar in two respects: 1st, no ligature was needed; and 2nd, at least two-thirds of the face of the stump was the surface left by the passage of the cannon ball, and yet it healed very kindly." Dr. Taylor informs me that he recently saw this man in good health. He is on the staff in Belfast.

favoured, all appear reasons against sutures. Strips of wet lint applied like adhesive plaster, always appeared preferable. I never saw one case among our most numerous amputations in which primary adhesion took place thoughout the whole surface of the flaps. They united readily enough along their edges; but the result of this was, that a large bag of pus was formed within the end of the stump, which continued as a depôt for absorption into the system, by steeping the end of the sawn bone and the vessels in its matter, and it burrowed far and wide in the intermuscular spaces and along the bone, and ended not unfrequently in causing considerable necrosis of the end of the divided shaft. Unquestionably it may be said that such collections should have been recognized and prevented; but yet it seems to me, that when ample proof is afforded, as it was early in the East, that primary adhesion was the rare exception, and not the rule ; and when the patients were so peculiarly liable to purulent absorption as they were with us: it would have been better practice not to have attempted primary union, but to have adopted such treatment as best favoured the freest discharge of the matter so soon as it was formed. The method of dressing with compresses, recommended by Mr. Luke, was most useful in several cases in which I tried it, in preventing the accumulations referred to. The contrast afforded by the heavy dressings for stumps employed by the French and our water dressing was very marked, and may have contributed something to the result which obtained in the less prevalence of purulent absorption in our hospitals than with them. Bad as it was with us, it never became the terrible epidemic it was in the French hospitals. We had no means of trying the method of treating stumps in water, recommended by Langenbeck. The ease with which the purulent secretion can be got quit of by position in amputations of the arm and leg, contributes, I have no doubt, not a little to the decreased mortality attending these operations, as compared to amputations of the thigh. The Russian surgeons, I am told, when operating by the circular method, which they generally adopt, split the posterior flap, and keep this part open in order to drain off the pus. Such a step would meet with little favour in this country, but it presents many advantages when purulent absorption is so common as it was in the East. M. Sedillot, of Strasburg, I believe, proposes a similar modification for general use.

Primary adhesion is, of course, most desirable when hospital gangrene prevails, but it is just at such a time that this result is most difficult to obtain.

Cases of secondary amputation of the thigh for injury of the knee, were always those in which attempts at primary union did worse. The long fusiform collections of matter which are so apt to exist in these cases previous to operation, extended, and did every possible harm. Careful bandaging from above downwards to the base of the flaps seemed to be highly useful in these cases.

Pus poisoning was unquestionably the chief source of our mortality in the East after amputation, especially after secondary operations. The resemblance between its early features and those of ague was perhaps more marked among our patients than it even usually is. This seemed especially the case among men who had served during the early part of the war—of this, however, I am not certain. We had many most beautiful examples, *post mortem*, of veins leading from the stump remaining round, patulous, and filled with pus, and sometimes reddened in their interior. It was not uncommon to trace the pus-filled vein from the thigh to the vena cava.

It is a question on which it is difficult to decide whether or not, when pus absorption is so common as it was with us, it would not be justifiable practice to ligature the chief vein at the time of operation. The views of Mr. Travers and others would certainly seem to oppose the adoption of such a step, but we have, on the other hand, the evident absorption of pus into the system by this channel; and, besides, numerous cases are on record in which the ligature of the vein has not only not been followed by evil results, but has absolutely been the apparent cause of preventing inflammation and pus absorption.* The non-appearance of symptoms of purulent poisoning till after the separation of the threads, makes it generally difficult to say which set of vessels—those ligatured or those not ligatured—have been the carriers of the pus. In the case referred to in the note death took place rapidly, before the ligatures were detached. Hennen expresses himself thus on the danger of tying veins: "When the great veins bleed I have never hesitated about tying them also, and it is most particularly necessary in debilitated subjects." Chevalier, too, says—"I know from experience that the principal vein of a limb may be included in the same ligature as the artery without any disadvantage ensuing." Every hospital surgeon has seen instances of the same thing. I most firmly believe in Stromeyer's views on absorption by the veins of the bone, from observations which have been presented to me.

Independently of all fortuitous circumstances, there can be little doubt but that some constitutions oppose themselves more to pus poisoning than others. This, although a most unsatisfactory mode of explanation, yet seems the only way of answering the difficulty which is presented to us in the much greater sus-

---

* This is particularly well illustrated in a case related by Mr. Johnston, of St. George's hospital, in the Journals of 1857. In that case those vessels which had been tied were free both of inflammation and pus, while those not included in ligatures were full of pus, and "much inflamed."

ceptibility of some to purulent absorption than others. Most die rapidly, while others, not apparently so well fitted to withstand the assaults of such an invader, though placed in precisely the same circumstances, only yield inch by inch, and others again escape altogether.

The presence of typhus fever in an hospital has been supposed to favour the development of pyœmia, and, although it cannot be denied but that the diseases often co-exist, yet it seems more probable that they both proceed from a like source—a lowered vital energy in the patients, or vitiated hygienic arrangements.

The secondary deposits were with us, as usual, generally found in the lungs. Beck states, as the results of his observation in Holstein, that such was the seat of the deposition in seven cases out of ten in which patients died of pyœmia. This is not, I believe, an exaggerated average. Some of the French surgeons employed at Constantinople made the remark that they seldom found the pus collected in depôts, as they had been accustomed to see it in Africa; but that it commonly was disseminated through the organs, muscles, and bones.

The visceral congestions which so often follow amputation, were more than commonly fatal in their results in the Crimea, from the presence in most cases of the seeds of disease in the lungs, kidneys, and intestines. Phthisis and acute dysenteric attacks were not unfrequently the immediate causes of death, and in at least two cases the symptoms of Bright's disease of the kidney were most rapidly developed after thigh amputations.

CONTENTS.

CHAPTER I.

Distinction between Surgery as practised in the Army, and in civil life—Soldiers as patients, and the character of the injuries to which they are liable—Some peculiarities in the wounds and injuries seen during the late War, - 5

CHAPTER II.

The "peculiarities" of Gun-shot wounds, and their general treatment, - - - - 14

CHAPTER III.

The use of Chloroform in the Crimea—Primary and Secondary hæmorrhage from Gun-shot wounds—Tetanus—Gangrene—Erysipelas—Frost-bite, - - - 31

CHAPTER IV.

Injuries of the Head, - - - - 57

CHAPTER V.

Wounds of the Face and Chest, - - - 84

CHAPTER VI.

Gun-shot wounds of the Abdomen and Bladder, - 104

CHAPTER VII.

Compound Fracture of the Extremities, · · 118

CHAPTER VIII.

Gun-shot wounds of Joints— Excision of Joints, &c., · 135

CHAPTER IX.

Amputation, · · · · 163

www.ingramcontent.com/pod-product-compliance
Lightning Source LLC
Chambersburg PA
CBHW020242170426
43202CB00008B/185